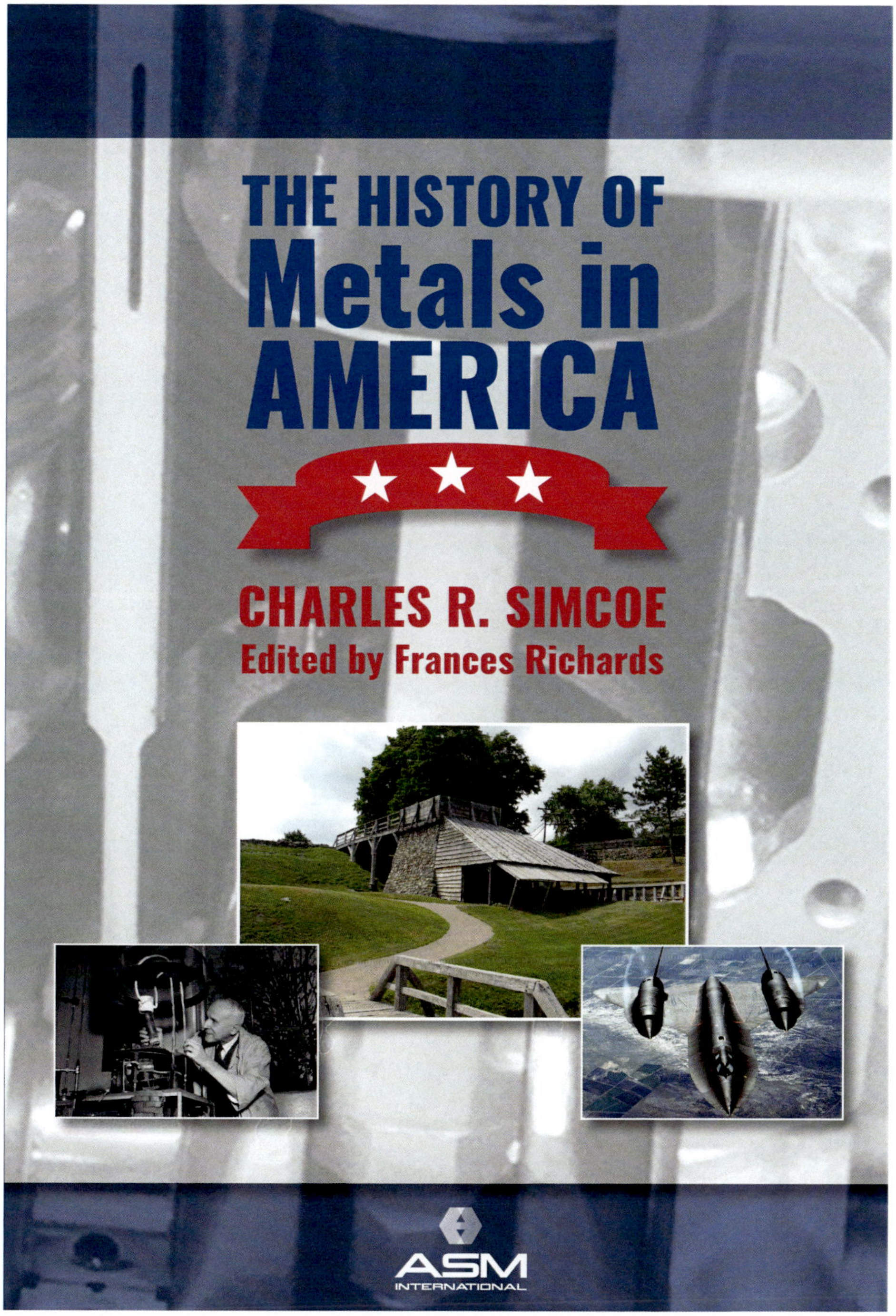

ASM International®
Materials Park, Ohio 44073-0002
www.asminternational.org

First printing, May 2018

Comments, criticisms, and suggestions are invited, and should be forwarded to ASM International.

Prepared under the direction of the ASM International Technical Book Committee (2017–2018), Craig Schroeder, Chair.

ASM International staff who worked on this project include Scott Henry, Senior Content Engineer; Karen Marken, Senior Managing Editor; Susan Sellers, Content Developer; Madrid Tramble, Manager of Production; and Jennifer Kelly, Production Specialist

Library of Congress Control Number: 2018933953

ISBN-13: 978-1-62708-145-0
ISBN-10: 1-62708-145-3
EISBN: 978-1-62708-146-7

SAN: 204-7586

ASM International®
Materials Park, OH 44073-0002
www.asminternational.org

Printed in the United States of America

To Mary
and our family Robert, Joyce, Thomas,
William, Richard, and Janice

Contents

Note to Readers

Becoming a published author was a longtime dream of our father, Charles Robert "Bob" Simcoe. As a metallurgical researcher by profession and a voracious reader of history and biographies, he wanted to pass his knowledge on to others. On February 25, 2017, however, he died quietly in the intensive care unit of Iroquois Memorial Hospital. He was just one month shy of turning 94.

During his final month, while in a rehabilitation center, he continued to revise and edit chapters of this book and began to write the preface. But the three paragraphs of the preface following this note are all he managed before his death. Those three paragraphs, though, can only hint at the story behind the scenes.

As a research metallurgist, our father worked to support a family of eight, with little time for extra activities. But he loved to tell stories about metals and would readily explain the structure of a building down to the tiniest screw or the composition of the soda can from which we might be drinking. After he retired, he began to put the stories in this book together, at first as a hobby and for his own amusement. He researched not only in libraries but also among his own files, along with interviewing friends in and retired from the industry and creating a blog for a wider audience.

Writing for long periods was interrupted by the need to care for our mother, Mary, who was diagnosed with memory loss and eventually Alzheimer's disease. As our mother's health deteriorated and she was placed in a nursing home, Dad again took time to write. Even though he spent countless hours with our mother, it was then that he began to contact editors at *Advanced Materials & Processes,* ASM International's flagship publication, about turning his blog into a series of articles. He did not let rejection stand in his way. On March 14, 2013, our mother passed away quietly in her sleep. By then, they had been married over 71 years.

In the following months, Frances Richards became the new editor of *Advanced Materials & Processes,* and Dad immediately pitched his writing ideas to her. She readily accepted them, and the "Metallurgy Lane" article series was born. Writing for the magazine and later this book became our father's purpose in life, his legacy. He came alive as he received emails from Karen Marken and

Frances Richards, his patient and understanding editors, and his readers, who hailed from all around the world. He raced life's clock to finish this book. We are sad that he will not be able to hold it in his hands, but grateful that he won that race.

Robert Simcoe
Joyce Simcoe Simutis
Thomas Simcoe
William Simcoe
Richard Simcoe
Janice Simcoe

April 2017

Preface

It was my intention, for many years, to write on the history of the development of modern technology from the viewpoint of the structural metals used to build our 21st century world. I started this work more than 30 years ago. Many times I have picked it up for a while and then laid it down to do more pressing things connected with earning a living. In addition, I have felt along the way that the subject matter is so vast and that other workers in the field were so much more qualified to perform the task that I was willing to defer to them. However, no one came forward to do the work. In the meantime, a whole generation of metals workers who might have written a more comprehensive history, a more knowledgeable history, indeed a more elegant history have passed from the scene.

So, I take up the challenge of presenting the history of the development of metals as both an industrial activity and a science, which made possible the present world of land, air, and space travel; of chemical production in rust-resistant plants; of buildings reaching over 100 stories high; of welded ships that can cruise for months at a time on nuclear power; and many other engineering accomplishments that the average reader takes for granted.

The past 100 years have been ones which indeed could be called a second industrial revolution, or more accurately a technological revolution. The industrial revolution was based on steam power. This technological revolution involves electricity, oil, and nuclear power combined with developments in transportation, communications, manufacturing advances, and general consumer needs. Nearly every major technical advance, however, has been accompanied or preceded by the development of a new metal or alloy or a breakthrough in the use of an existing one.

About the Author

Charles Robert "Bob" Simcoe became a member of ASM International, formerly the American Society for Metals, in 1950. He attended The Ohio State University and graduated from Purdue University with a degree in metallurgical engineering in 1950. Simcoe became interested in the field while serving in the U.S. Navy during World War II and taking a course in welding and metallurgy during his service.

After graduation, his first job was with Westinghouse Atomic Power Division in Pittsburgh, where he studied zirconium, the structural metal for the atomic reactor in the USS *Nautilus* submarine. After two years, he began work at Battelle Memorial Institute in Columbus, Ohio. He focused on alloy steels and hydrogen in steel and titanium. In 1958, he began work for the Armour Research Foundation in Chicago, Illinois. He directed a team working on titanium, aluminum, alloy steels, columbium, and other alloys. In 1964, he moved to Lockport, New York, to work for Simonds Saw and Steel. It was here that he worked as an assistant laboratory director, materials manager, and vice president of sales and marketing until he retired in 1985.

During his retirement, he consulted with various businesses and worked for Curtiss-Wright in Buffalo, New York. He also taught the metallurgy lab at the State University of New York at Buffalo for six years.

During his career, Simcoe wrote articles for *Transactions of the ASM*, *The National Metalworking Weekly*, a publication of *Centre D'Information Du Cobalt* (Brussels), *Mechanical Engineering*, and the *Journal of Metals*. He also wrote more than 40 articles for ASM International's *Advanced Materials & Processes* magazine, which became the basis for this book.

Chapter One

The Discovery of Metals

The history of metals in America started in antiquity when man discovered particles of gold, silver, and copper in riverbeds where they had washed down from the mountains. These particles of soft metals were hammered into objects for jewelry, masks, and religious worship. Early humans had no knowledge of iron. Unlike gold, silver, and copper—known as native metals because they exist in their elemental form uncombined with any other elements—iron does not exist as a native metal. All iron, except for occasional meteorites, is tied up in compound form. Therefore, its discovery was fortuitous; humans built fires with wood, placed stones within the fire pits to maintain heat, and learned that iron was reduced from these iron-rich stones. This type of iron was not melted but was instead formed in a solid state as stone and slagged off, leaving a mass of iron particles at the bottom of the pit.

This iron was essentially carbon-free because it never reached a temperature in which carbon could be absorbed. With time, special fire pits were built to increase the amount of iron that could be produced. Bellows, a device used for blowing air into a fire, were used to raise the temperature. With further refinements, this process was used for many centuries to produce what we call "*wrought iron*." During the many early years that humans used iron, wrought iron was the dominant form. It could be worked hot or cold to produce weapons, armor, knives, and numerous items for everyday life. One of the most interesting objects made with wrought iron is the Ashoka Pillar in New Delhi, India (**FIG. 1.1**). It is at least 1600 years old and is 16 inches in diameter, 24 feet high, and weighs approximately seven tons. It is a remarkable work of metalsmithing for the ancient times.

The use of wrought iron increased during the Middle Ages. For example, it was used to reinforce joints in the new water wheels and windmills. Armies also used increasing amounts in armor and weapons in multiple wars in the 13th and 14th centuries (**FIG. 1.2**). By the Middle Ages, blowing engines and larger furnaces increased temperatures until the iron melted. As recorded by R.F. Tylecote in his book *A History of Metallurgy*, "the development and introduction of the blast

FIG. 1.1 The Ashoka Pillar in New Delhi, India. This wrought-iron artifact has resisted corrosion for at least 1600 years.

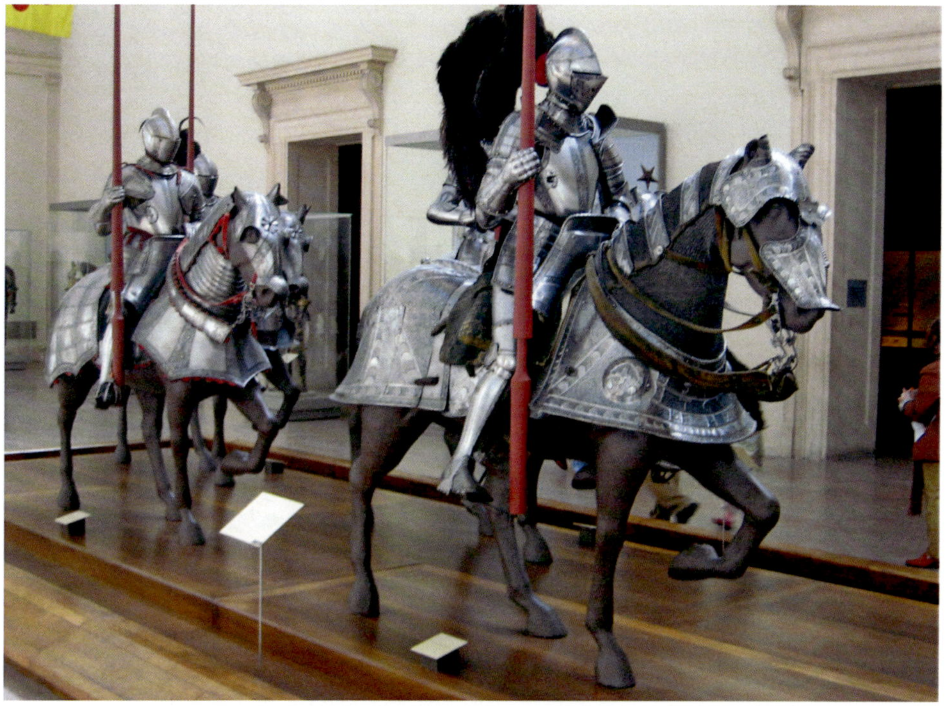

FIG. 1.2 These mounted knights showcase the largest use of wrought-iron sheet during the Middle Ages. Source: Wikimedia Commons/Mattes.

furnace in Europe is one of the most interesting subjects in the history of ferrous metallurgy," (Ref 1). The molten iron dissolved the carbon from the wood fuel until the carbon reached 3–4% of the iron. When this iron cooled, it was brittle and could not be worked either hot or cold and could only be cast into molds that formed the shape of the finished article. This *cast iron* could be used for cannons, cannon balls, utensils, and eventually stoves, lampposts, and other products, but it could not replace wrought iron.

For hundreds of years, wood and charcoal were the major fuel for all industrial processes—shipbuilding, brickmaking, glassmaking, pottery, and even brewing. But a truly prodigious use of charcoal was in ironmaking. Charcoal not only produced the heat for the process but also provided the element (carbon) in the chemical reaction that reduced the iron ore to iron. Blast furnaces were built wherever iron ore was found, and woodchoppers would begin to strip the surrounding countryside of all standing timber. Roughly one acre of woods was needed for each ton of iron produced. Ironmaking in an area stopped only when wood to fuel the furnaces was gone. By the beginning of the 1700s, this created a serious shortage of timber in England, Scotland, and Wales, as well as in countries on the European Continent. As a result, ironmaking had started to decline in England long before the Industrial Revolution. England thus became an importer of iron.

The problem was solved by the increasing use of coke, the residue of coal, for iron smelting from about 1720 onward. For metallurgists, this marks the beginning of the Industrial Revolution. Abraham Darby and his partners were the first to achieve this in a practical way.

Abraham Darby came from a family of farmers. The Darbys were Quakers, a religious group that seemed to take naturally to business and industry. In his youth, Darby was apprenticed to a maker of malt mills, brass equipment for brewing beer. He visited Holland in 1704, where he saw coke being used as the fuel in a variety of industries, including brass foundries. Several years later, he entered the brass business in Bristol, where he used coke for fuel. In 1707, he and John Thomas, another Quaker, took out a patent for manufacturing iron products. The next year, he founded the Bristol Iron Company to make cast-iron products in competition with brass.

In 1709, Darby moved his ironworks up the Severn River from Bristol to Coalbrookdale, where he leased an old furnace (circa 1638) that had been abandoned by the previous operators. He used coke in making cast iron to manufacture pots and other wares of the local trade. He died in 1717, without the general ironmaking industry being aware of his great technical achievement (**FIG. 1.3**).

Ironmaking continued at the Darby Works under Richard Ford, a son-in-law, until it passed on to Abraham Darby II around 1732. It was long believed that Darby II may have had more to do with the successful use of coke than his father. In his metallurgy book, Tylecote wrote, "the first change made by Darby II was the use of the Newcomen steam engine," (Ref 2). It seems reasonable that the longer-range development of larger and taller blast furnaces and the greatly increased blowing power along with an efficient balance of iron ore, coke, and limestone in the charge would have come gradually over time with careful experimentation on the part of the ironmasters. Following Abraham Darby II, who died in 1763, the firm was again managed by a son-in-law, William Reynolds, prior to the ascendance of Abraham Darby III. Historians have seriously questioned whether or not England could have achieved its success in the Industrial Revolution without being, at the same time, a successful metalworking nation. We will never know because the Darbys and the little iron bridge at Coalbrookdale thrust the country into the lead as the world's maker of iron (**FIG. 1.4**).

By the 1770s, the use of coke in ironmaking was becoming more common, though not universal. Ironmasters other than the Darbys were becoming known for the quality of their products and for their astute business sense. Foremost

FIG. 1.3 The Darby Furnace was enlarged in 1777 to build the cast-iron bridge at Coalbrookdale. Source: Wikimedia Commons/Helen Simonsson.

among these outstanding masters was John Wilkinson, also of Coalbrookdale. Wilkinson's father, Isaac, may have been the first ironmaster to follow in the footsteps of the Darbys in the use of coke. John Wilkinson was the first ironmaster to install the new steam engine built by James Watt. Wilkinson became a cast iron crusader. He believed that cast iron literally could be used for everything. He developed cast-iron pipe and was reported to have supplied over 40 miles of it for water mains for the city of Paris in 1788. He bolted cast-iron plates together to build a boat, which he used on the Severn River to the wonderment of local townspeople. They found it incredible that cast iron could float. He carried his enthusiasm for cast iron to his grave and was buried in a cast-iron coffin. About the time the Revolutionary War was starting in the American colonies, Abraham Darby III and John Wilkinson proposed that the new bridge across the Severn at Coalbrookdale be made of cast iron. There was no precedent for using iron on such a scale and for such an important and costly structure. The bridge was designed by a local architect by the name of Thomas Pritchard, who died before the bridge was built. The design was considerably altered during construction, which took place between 1775 and 1779 (**FIG. 1.5**). Barrie Trinder, a historian of the Industrial Revolution in Shropshire, wrote in his book, *The Darbys of Coalbrookdale*, "Pritchard was an architect, a stonemason by training, and not an ironmaster, and there can be little doubt that the structure as it was finally realized was to a large extent determined by Abraham Darby III and his foundry men."

FIG. 1.4 Ironmaking at night at Coalbrookdale. Painting by Philipp Jakob Loutherbourg.

The metal was cast in the Darby Iron Works because Darby's shop was close to the construction site. In fact, it is believed that the "old furnace" leased by Abraham Darby in 1709 was enlarged to increase its capacity for this project. Thus the greatest iron dynasty of 18th-century England and the development of coke to replace charcoal in making high-quality cast iron brought forth one of those rare engineering accomplishments—a near-perfect structure that was both useful and beautiful. No massive parts, no overdesign to allow for doubt or ignorance, no excess material whatsoever. It was the marriage of a new material with a design reflecting a growing awareness of engineering possibilities. This little iron bridge is the ancestor of all present-day metal structures. Trinder again wrote in *The Darbys of Coalbrookdale*, "the Iron Bridge was no mere milestone in the history of civil engineering: It was the phenomenon of the age." In 1788, the Society of Arts awarded Abraham Darby III its gold medal for the magnitude and importance of the iron bridge.

Technical advances came slowly 200 years ago. It was nearly a generation before another major iron structure was built. Others rapidly followed. A second bridge nearly twice the span of the Coalbrookdale Bridge was constructed at Sunderland over the Wear River about 1795. The great Scottish engineer Thomas Telford incorporated cast iron bridges and aqueducts into his roads and canals. Within 25 years after the construction of the bridge at Coalbrookdale, Telford completed his greatest achievement in canal building by carrying the Llangollen

Canal in a cast-iron aqueduct over the River Dee in Wales for a span of 1000 feet at a height of 127 feet (**FIG. 1.6**). Cast iron continued to be used for major structures as late as 1850 when Robert Stephenson, the railroad builder, spanned the Tyne River at Newcastle.

One of the great accomplishments in bridge building was the Menai Strait Bridge designed by Thomas Telford and constructed in 1826 (**FIG. 1.7**). The structure is entirely made of wrought-iron bars connected to form a chain for the suspension bridge. At the time, it was the longest suspension bridge in the world at 1000 feet long. It is considered Telford's greatest engineering accomplishment.

The closing years of the 18th century and the opening ones of the 19th provided great opportunities for the ironmasters of England. The Darbys supplied many of the cast-iron cylinders for steam engines as early as 1723 for the Newcomen-type engines and after about 1785 for the new Watt engines. They made the cast-iron beams and posts for the first fireproof textile building in 1797. The Darbys and their Coalbrookdale Company also made the cast-iron gates of the

FIG. 1.5 The Iron Bridge at Coalbrookdale, built in 1779, is the first structure made of cast iron. Painting by William Williams.

FIG. 1.6 The cast-iron aqueduct that carries the Llangollen Canal over the River Dee was built in 1805 by Thomas Telford.

famous Crystal Palace Exhibition Hall in London in 1851 (**FIG. 1.8**). These gates are still in use today between Hyde Park and Kensington Gardens in London.

The last of the Darbys to be associated with the Coalbrookdale Company were Alfred Darby and Abraham Darby IV, who both retired in 1849, and Francis Darby, son of Abraham Darby III, who died in 1850. It was Francis Darby who supplied the artistic talent for making art castings that the company became known for in the 1840s. Thus, the Darby ironmasters spanned the period in Shropshire from 1709 to 1850, over 140 years. During most of these years the family had many partners. They seldom held complete control but they always made significant contributions.

The iron trade declined in the Shropshire region throughout the 19th century, especially after 1850. Near the end (1871), a committee of the Iron and Steel Institute, including Henry Bessemer, visited to look for old documents and for surviving relics of Abraham Darby I. Cast iron became low enough in cost to find widespread use. It was cast into such common products as posts for roadway

FIG. 1.7 The wrought-iron Menai Suspension Bridge in the United Kingdom was designed by Thomas Telford in 1826. Source: Wikimedia Commons/Bencherlite.

lamps and signs, church stoves, complex window frames, and even tombstones (reminiscent of John Wilkinson). It became fashionable by the mid-19th century to build the street side of public buildings and even private homes with cast iron for structural supports and decorative trim.

The Darbys, John Wilkinson, and William Reynolds (the same Reynolds who managed the Darby Works as a son-in-law) of Coalbrookdale, along with other great ironmasters such as the Crawshays of Cyfarthfa, Wales, and John Roebuck of the Carron Iron Works in the lowlands of Scotland, all laid the groundwork for the important metalworking segment of the coming Industrial Revolution. They and some of their descendants would continue to improve the technology and economics of basic ironmaking.

One of their colleagues, Henry Cort, developed the "puddling process" in 1784 to convert cast iron from the blast furnace into wrought iron. This puddling technique was the most important development in metals for the coming 19th century. The process was carried out in a reverberatory furnace that included cast iron rather than iron ore. Heat and gases from the fire passed over the cast iron, both melting it and removing the carbon. Cast iron melts at roughly 2100 °F

FIG. 1.8 The Coalbrookdale Gates, built at the Darby iron plant for the Crystal Palace Exhibition of 1851, now reside at Kensington Gardens in London. Source: www.royalparks.org.uk.

and is dissolved into the slag. As carbon is removed, the melting point of the carbon-free iron increases to 2800 °F, higher than the furnace temperature. The iron then precipitates as solid particles that fall through the slag and onto the hearth as a mixture of iron and slag. This mixture is then agitated with a paddle to form a semisolid ball, which is handled as ordinary wrought iron and made in batches of approximately 500–600 lb. Excess slag is squeezed out, leaving an iron containing just 2–3% slag.

Wrought iron was the major metal for the high period of the Industrial Revolution in the 19th century. It enabled the use of power on a massive scale for trains, ships, steam engines, bridges, and machinery. The use of coke, the puddling process, and the Bolton and Watt steam engine, invented by James Watt at the University of Glasgow and in use by 1775, allowed Britain to reclaim her lead in ironmaking. From early in the 18th century until the beginning of the 19th century, ironmaking increased from 20,000 to 250,000 tons per year.

Cast iron is no longer used in structures such as bridges and buildings, and wrought iron is no longer produced. Nevertheless, our modern world of 100-story skyscrapers, superhighways, suspension bridges, and many other structures can be traced directly to the Abraham Darbys and the Iron Bridge at Coalbrookdale.

REFERENCES

1. R.F. Tylecote, *A History of Metallurgy*, 2nd ed., London: Maney Publishing, 1976
2. B.S. Trinder, *The Darbys of Coalbrookdale*, Phillimore & Co. Ltd., 1992

Selected References

B. Bracegirdle, *The Archaeology of the Industrial Revolution*, Heinemann, 1973

J.L. Bray, *Ferrous Production Metallurgy*, John Wiley & Sons, 1942

A. Feldman and P. Ford, *Scientists and Inventors: The People Who Made Technology from Earliest Times to Present Day*, Facts on File, 1979

Chapter Two

Iron in America—1645 to 1870

Ironmaking commenced in the colonies as early as 1645 at the Saugus Iron Works in Massachusetts. This effort was never financially successful and closed in 1676. The site was completely restored as part of America's industrial heritage (**FIG. 2.1**). A more successful metal operation was the Raynham Forge Plant near Taunton, Massachusetts, which opened in 1656 and operated until 1880. Other furnaces were built at Braintree, Massachusetts, in 1648 and at New Haven, Connecticut, in 1658. Thomas Rutter built the first iron works in Pennsylvania with a bloomery forge in Berks County in 1716 and the first blast furnace, the Colebrookdale Furnace, in 1720. Anthony Morris built the famous Durham Iron Works near Easton, Pennsylvania, on the Delaware River. Peter Grubb discovered the iron ore deposit where he located the famous Cornwall Mine in 1736. This open pit mine became the largest source of iron ore in America until the opening of the great ore deposits of Michigan and Minnesota more than a century later.

Ironmaking Spreads to More Colonies

Between the development of these very early iron plants and the time of the American Revolutionary War, ironmaking spread to many of the colonies in the northeast. Wherever bog iron ore or other iron-bearing minerals were found in conjunction with plentiful supplies of wood for charcoal, iron was made, provided either a local population or water transportation to a ready market was also present. Ironmaking spread to New Jersey by 1710, with other operations

FIG. 2.1 The restored Saugus Iron Works in Massachusetts. Source: Wikimedia Commons/Daderot.

extending to the Oxford Furnace (1742–1882) and Martha Furnace in the Pine Barrens of southern New Jersey. The Martha Furnace was a 20,000-acre plantation, one of the largest ever operated. The Long Pond Ironworks near West Milford, New Jersey, started by Peter Hasenclever, was one of the largest investments in an iron works prior to the Revolutionary War. The investment totaled approximately $250,000. It remained in operation for 120 years until the new iron and steel works in Pittsburgh made it obsolete in 1880.

Peter Grubb's Cornwall Furnace was built at his mine in 1742 and remained in continuous operation until 1883 (**FIG. 2.2**). The furnace, along with its associated buildings and the town that was built for the Welsh workers imported to work at the furnace, is the most intact original iron works in the United States, at nearly 275 years old.

The Cornwall Furnace remained in the Grubb family until 1798, when it was acquired by an iron company controlled by Robert Coleman, who owned a number of furnaces and forges in the area. Coleman became wealthy with his many investments and was reported to be the first millionaire in Pennsylvania. The Cornwall Iron Works remained in operation under the Coleman family for 85 years until 1883, when it closed due to competition from the new Bessemer

FIG. 2.2 The original iron works built by Peter Grubb in 1742. Source: Wherring at en.wikipedia.

plants in Pittsburgh, Steelton, and Bethlehem, Pennsylvania. The era of steel had arrived. The Cornwall mine remained in operation and was purchased in 1920 by the Bethlehem Steel Company. The underground mine was closed when it flooded during Hurricane Agnes in 1972, and during the following year, the open pit mine ceased operation. The Coleman family retained ownership of the property until 1932, when it was given to the Commonwealth of Pennsylvania as a historical site.

The Principio Plant in Maryland was another large operation. In 1751, the original facility included four blast furnaces and two forges in addition to timber covering 30,000 acres (**FIG. 2.3**). The plant was destroyed by the British Army in 1777 and rebuilt during the War of 1812, but the British destroyed it a second time. The Whitaker brothers acquired it in 1836 and restored it again. The Principio Company was absorbed into Wheeling Steel Corp. in 1920.

Pennsylvania became the major iron-producing region by the time of the Revolutionary War. The availability of ore, wood for charcoal, limestone for flux, and water from the Susquehanna, Schuylkill, and Lehigh-Delaware Rivers and their tributaries kept the region productive until wood for charcoal became scarce, as happened earlier in England.

FIG. 2.3 Office of Principio Iron Works, the company's only remaining structure.

As John L. Bray points out in his textbook on *Ferrous Production Metallurgy* (1942), "The story of the great 'iron plantations' which flourished in Pennsylvania during the eighteenth century is one of the least known but most interesting chapters of colonial history," (Ref 1). These plantations (up to 30,000 acres) usually contained a large mansion for the owner that was surrounded by houses for 60–80 workers and their families. The greatest activity was cutting wood and making charcoal because the blast furnace consumed the harvest from an acre of land on a daily basis. Iron ore was mined from the local hillside and delivered to the blast furnace by wagon or even in pouches on horseback. Output from the furnace as cast iron, or from the forge as wrought iron, had to be delivered to the outside world in the same manner. Yearly production of these iron plantations ranged from a few hundred tons to as much as 1000 tons for the largest facilities. In this fashion, the ironmakers of colonial America turned out approximately 30,000 tons of iron in 1775.

One of the largest projects during the Revolutionary War was a massive wrought-iron chain installed across the Hudson River at a narrow bend at West Point (**FIG. 2.4**) with the purpose of preventing British ship passage on the river. The chain was made of 750 links. Individual links were hammered out, flattened at the ends, bent around a mandrel, and hammer-welded at the joint. The chain

FIG. 2.4 Portion of the wrought-iron chain that was installed across the Hudson River at West Point.

was made up of sections containing nine links each that were floated on rafts and attached into a continuous chain with clevis joints and pins during installation. The completed chain was anchored at both sides of the river, and the rafts carrying the sections remained in place to support the weight. The British never tested the installation. The Hudson above West Point was strictly a Colonial waterway, and the flow of men and supplies from New England to Virginia was assured.

Iron production in the new United States of America increased from 30,000 tons at the start of the Revolution (1775) to 54,000 tons from 153 furnaces by the census of 1810. This slow growth was largely due to a general lack of industrialization in America; per capita consumption of iron hardly changed in this 35-year span. Another factor was imported iron from England. It was cheaper to make iron from coke in Britain and transport it by water to the major port cities of the East Coast than to continue using charcoal in the inland plantations of America and transport product to the newly growing population centers.

The conflict with England during the War of 1812 increased demand for iron but did little to change the nature of making it. The United States continued to use charcoal as fuel for reducing iron ore in the blast furnace. This was a matter not of backward technology but of availability. Massive deposits of bituminous coal lay in the Pittsburgh area, across the mountains from where the iron furnaces were built in southeastern Pennsylvania. Even the anthracite coal discovered in northeastern Pennsylvania could not reach the iron furnaces. It is difficult today to visualize the extreme handicap imposed on the United States during this period before the advent of bulk transportation.

Canal Building in Pennsylvania

The beginning of a solution to this problem came about when the city of Philadelphia was desperately seeking a way to prevent farm and manufactured goods, produced in areas north and west of the city, from using the Susquehanna River to ship from Baltimore. Philadelphia was the largest port in this new country, and its politicians were battling to maintain its position against New York City and nearby Baltimore. After 1825 and the Erie Canal opening, New York was clearly in a position to dominate the trade coming from the upper Midwest. This left Philadelphia with the problem of finding a practical route through several mountain groups making up the Allegheníes to Pittsburgh, where trade could funnel through the Ohio River into the center of the nation.

The first priority was to capture the local trade leaking out to Baltimore and provide a way to bring coal from the newly developed anthracite coal mines in the northeastern corner of the state to the metropolitan areas along the coast. The only solution to bulk transportation at this point in history was by water travel. While there were many natural streams in this corner of Pennsylvania, they all contained numerous obstacles to navigation. This meant that canals needed to be built. Canals were not new to America. They had been dug in New England with limited success. One successful canal ran across upstate New York to Lake Erie, a job just coming to completion in the mid-1820s. The first canal built in America, and one of the most profitable, was the Delaware and Hudson Canal (D&H Canal). It was built by private capital (many of America's early canals were built by private companies; state governments sometimes helped by giving these companies the right to revenue from lotteries). The canal was masterminded by three brothers by the name of Wurts, but they had two major problems. The first was to find a way to get the coal out of the wilderness of northeast Pennsylvania and down to cities on the coast, especially New York. The second was to convince people that anthracite coal (known as stone coal) was an acceptable fuel. Anthracite coal was formed under more heat and pressure than the better-known bituminous coal. Thus, it was harder and lower in volatiles, and it required a higher temperature to ignite and burn. The Wurts brothers felt that if they could get the coal to New York at a reasonable cost, its value would soon sell itself. The mayor of New York City, Philip Hone, was their ally in this endeavor.

FIG. 2.5 Benjamin Wright, chief engineer of the Erie Canal, was hired to build the D&H Canal.

Technical expertise for the canal project came from Benjamin Wright, who had been chief engineer on the Erie Canal (**FIG. 2.5**). The logical route for the canal, recognized even by the Wurts brothers, was across the Delaware River into the state of New York, northeast up the valleys west of the Shawangunk Mountains, and then down Rondout Creek to Kingston and the Hudson River. It was a natural waterway even though it meant the canal went northeast when New York City was southeast. Due to this configuration, the longest mileage on the haul was on the Hudson River.

Work began in the summer of 1825 when the Erie Canal's five locks at Lockport, New York, were completed. This released Wright and several of his assistants from Erie to work on the Delaware and Hudson. The 60-mile section within New York was completed in just over a year, but the more challenging stretch up the Delaware and inland to a little village called Honesdale, Pennsylvania (named for the mayor of New York City), required an additional year and a half. While the canal officially opened in fall 1828, and some coal was immediately shipped to New York City, major projects remained. One was the incline required to bring coal over the 1000-foot Moosic Mountain to Honesdale and the aqueduct to carry the canal across the Delaware River.

The first project was a system of inclines to carry the coal from the mines over Moosic Mountain down to the canal's terminus. For part of this route, a railroad was constructed, and a young engineer by the name of Horatio Allen was sent to England to buy several locomotives. One of these, the Stourbridge Lion, was actually used in a trial run and had the honor of being the first steam locomotive operated in the United States. However, at eight tons in weight, it was obvious on its first and only run that it was too heavy for the light rails and bridges. It was never used again, and years later the canal company donated it to the Smithsonian Institution. The inclined plane system for lifting the coal over the mountain was accomplished in a few years, and with the building of Roebling's aqueduct over the Delaware River in the late 1840s, the Delaware and Hudson Canal was finally completed (**FIG. 2.6**). It was one of the most prosperous transportation systems of the 19th century. Its major haul would always be anthracite coal, with as much as 3,000,000 tons in its peak year. It was finally abandoned in 1899, but the company name would continue as a railroad into the 1980s.

Other canals were built on the Susquehanna, Schuylkill, Lehigh, and Delaware Rivers. The Union Canal system linked the Susquehanna and Schuylkill with the coal fields above Reading. The big project finally linking the Susquehanna with the far distant Pittsburgh was the Main Line Canal, the most ambitious canal project in America, built at a cost of over 10 million dollars. It started at the Susquehanna and went up the Juniata River to Hollidaysburg at the foot of the

FIG. 2.6 The Roebling Aqueduct across the Delaware River. Source: Wikimedia Commons/Derek Ramsey.

Alleghenies. A series of inclined planes and rail lines carried it over the 37-mile mountain range down to Johnstown. From there it followed the Conemaugh River to the Allegheny River and then to Pittsburgh and the Ohio River. Philadelphia finally had a transportation system into the heart of the Midwest.

Anthracite Coal Replaces Charcoal in Ironmaking

With the sudden availability of anthracite coal in southeastern Pennsylvania, the ironmasters soon learned to adapt it to ironmaking. The hard coal did not have to be processed into coke; it was used as it came from the mine. Quoting from Robert B. Gordon in his book *American Iron*, "Ironmasters using anthracite fuel reduced the labor cost for smelting to $2.50 per ton at a time when the proprietors of the Hopewell charcoal-fired furnace paid wages amounting to $15.14 per ton," (Ref 2). With the use of anthracite coal for fuel, steam power to replace water power, and canals for transportation, ironmaking moved from isolated rural areas into more populated areas. The location and scale of ironmaking was undergoing rapid changes, and by the middle of the 19th century, more iron was made with anthracite coal than from charcoal.

Just as coke had revived the British iron industry a century before, anthracite coal accomplished the same result for the iron industry in southeast Pennsylvania. The industry had been in decline because it too was running out of wood, permitting the southeastern corner of Pennsylvania to become the first big iron producer in America. It was many years before Pittsburgh replaced Philadelphia as the biggest iron-producing region. During this time, Philadelphia was producing the consumer goods, mill machinery, and factories for manufacturing products made of iron (**FIG. 2.7**) and was thus a major incubator for the American Industrial Revolution.

Ironmaking Pioneer John Fritz

One distinguished name during this period of ironmaking was that of John Fritz (**FIG. 2.8**). Fritz was born on a farm in Eastern Pennsylvania in 1822, and at the age of 16 was apprenticed to the blacksmith trade. In 1842, he went to work as a common laborer in a new mill being built in Norristown, Pennsylvania, which made wrought iron and used rolling machinery for forming. When the mill went into production, Fritz was put in charge of the machinery and later was made manager of production. In the seven years he worked at Norristown, Fritz became experienced with all aspects of puddling and rolling wrought iron, especially the repair and improvement of machinery. Fritz moved to Safe Harbor, Pennsylvania, on the Susquehanna River in 1849 where a new mill was being built that included a blast furnace. This mill was built under Fritz's supervision, and during the first year of operation he resolved the startup problems.

FIG. 2.7 Fireplaces such as this Franklin stove were among the many household items made of cast iron.

John Fritz's next move was to another new mill in Johnstown, Pennsylvania. This mill was called Wood, Morrell and Company, but it would later become well known as the Cambria Iron Company. An entrepreneur by the name of Daniel Morrell leased a not-yet-completed new facility in Johnstown, where two previous sets of investors had gone bankrupt trying to build an integrated mill for producing railroad rails. This mill used coke in its blast furnaces and contained puddling furnaces and rolling mills. Fritz's first job, one he was now quite accomplished at, was to get all the machinery to operate properly and coordinate the flow of iron from the blast furnace to puddling furnaces to squeezers and finally to the rolling mills. He found that the local ores were not suitable for the quality needed to make rails. He had to convince the already financially strapped principals to buy better ore and pay a higher freight charge to bring it to Johnstown. But this did not solve the problem of excessive scrap during rolling. The mill was losing money, and Fritz decided that the current manner of rolling would never allow a solution to the problem.

The standard method of rolling at this time was to pass the hot billet through the rolls, and with pivoted tables on the output side, the semifinished rail was lifted up and shoved back across the top roll to the entrance side where it was again passed through a second groove in the rolls to further reduce the metal to a rail shape and size. The process was repeated in successively smaller openings until the final rail was formed. Fritz concluded that because the rails were cracking in the later stages of rolling, the metal was too cold for the attempted reduction. His solution was to stack a third roll over the two working rolls with a second set of grooves between the second and third rolls so that the rail could

FIG. 2.8 Paul Hercher's 1898 oil painting of John Fritz, inventor of the three-high rolling mill for producing railroad rails. Source: Linderman Library, Lehigh University.

be reduced in both directions. This helped keep the metal hot by the faster reduction and shortened total rolling time. Fritz designed and built the first three-high rolling mill in just 26 days. It was immediately successful, and soon all rails manufactured in the United States were made using his three-high mill design.

However, with the task of starting a new plant while solving numerous mechanical and metallurgical problems and always battling a lack of adequate finances, Fritz left Cambria Iron. He wrote in his notes, "After six years of as hard, laborious, faithful, vexatious work as ever fell to the lot of man to do, I decided to leave the scene of my early struggles and try my fortunes elsewhere." The elsewhere happened to be the new Bethlehem Iron Company, where he would work for the remainder of his long career. He was rewarded years later by the four major engineering societies issuing a medal in his name. The John Fritz Medal is one of the most prestigious honors in engineering to this day.

Ironmaking Expands for the Railroads

In the first half of the 19th century, iron production increased rapidly, from about 50,000 tons at the beginning to over 500,000 tons by 1850. The demand for this quantity of iron came from the railroads. Railroading started in the United States in 1830 with about 10 miles of track outside of Charleston, South

Carolina, and somewhat more laid by the Baltimore and Ohio Company. By 1840, there were 3000 miles of track and 9000 miles by 1850. During the 1850s, another 20,000 miles were added to give the United States 50% of the world's total, and this was before a single rail was laid west of the Mississippi. All of this tonnage of railroad rails was rolled from wrought iron.

As the population was moving west, the demand for iron kept moving west as well. In 1850, Chicago had one rail line and a population of 29,000; by 1860, it had 11 rail lines and 109,000 people. Demand for iron caused a burst of iron-making west of the Allegheny Mountains in Pennsylvania and into Kentucky and Ohio. Southeastern Ohio developed over 50 blast furnaces that supplied iron through the Civil War. This iron was still made using charcoal for fuel, except for the mills developing around Pittsburgh, which used coke made from bituminous coal located in nearby mines. While the production of wrought iron increased during the last 50 years of the 19th century, wrought iron was replaced after 1875 by the newly developed Bessemer process for turning cast iron from the blast furnace into steel.

REFERENCES

1. J.L. Bray, *Non-Ferrous Production Metallurgy*, John Wiley & Sons, 1941
2. R.B. Gordon, *American Iron: 1607–1900*. The Johns Hopkins University Press, 1996

Selected References

A. Bartholomew (Comp.) and L.E. Metz (Res.), *Delaware and Lehigh Canals*, Center for Canal History and Technology, Hugh Moore Historic Park and Museums, Easton, PA, 1989

R. Raymond, *Out of the Fiery Furnace: The Impact of Metals on the History of Mankind*, The Pennsylvania State University Press, 1986

B.C. Stoddard, *Steel: From Mine to Mill, The Metal that Made America*, Zenith Press, 2015

G. Tweedale, *Sheffield Steel and America: A Century of Commercial and Technological Interdependence 1830–1930*, Cambridge University Press, 1987

R.F. Tylecote, *A History of Metallurgy*, 2nd ed., Maney Publishing, 1976

R.F. Weigley, *Philadelphia: A 300 Year History*, W.W. Norton & Co., 1982

Chapter Three

The Age of Steel—1870 to 1900

The industrial world underwent a revolution in the last half of the 19th century. A process for making steel directly from blast-furnace cast iron was being developed at two different locations. One was at a small iron mill in Kentucky owned by two brothers by the names of William and John Kelly. It was typical for families to own and operate these small mills. William Kelly's early career was in wholesale dry goods in his hometown of Pittsburgh. He traveled widely in Ohio, Indiana, and Kentucky to take orders and collect accounts. It was during these travels that he met a young lady who lived in Eddyville, Kentucky. They married, and Kelly moved to Eddyville, where he looked for a business opportunity that he could buy and operate. He purchased, along with his brother, an iron plant with a large tract of timber for making charcoal. As usual, they soon had a fuel problem after clearing all the timber near the plant. During this time, William Kelly observed that where the air blast impinged on the molten metal in the finery, the metal was whiter in color or hotter; the cold air was not chilling the metal but was reacting with it chemically to generate heat. Kelly immediately saw the value of this reaction as a means of converting pig iron to wrought iron (**FIG. 3.1**).

Experimental Work Begins on "Air Blowing" to Make Steel

Kelly began experimental work in 1847, but he was delayed in building a converter until 1851. His intense interest in his experiments interfered with his attention to his regular iron trade, so his financial backers, including his father-in-law, finally

FIG. 3.1 William Kelly, American inventor of the "air-boiling process" for making steel from cast iron.

gave him an ultimatum to tend to business and stop his foolish ideas on the "air-boiling process," as they called it. In fact, when they found that he was secretly experimenting outside of his plant, his father-in-law had Kelly examined by a physician, Dr. Alfred Champion. Apparently Kelly communicated much better with this technically trained man than with his family, backers, and neighbors because he convinced Champion of the merits of his experimental process. Later, Champion served as a witness for Kelly during an interference ruling at the U.S. Patent Office on an American patent application.

The second location of experimental work on making steel from cast iron was in England under Henry Bessemer (**FIG. 3.2**). Bessemer was an experienced inventor with a substantial income from his patents, which included manufacturing bronze powder, devising machinery for extracting juice from sugar cane, and creating a rifled cannon shell that rotated during flight to improve the accuracy of the round cannonball in common use. It was during a demonstration in France where the cast-iron cannon ruptured when fired with his shell that Bessemer started a search for an improved metal for cannons. He knew that wrought iron or steel would be suitable, but there was no process to produce these in sufficient mass to make a cannon. Wrought iron and low-carbon steel both had melting points well above the temperature capabilities of available furnaces.

Somewhere in his search, Bessemer learned that a blast of air onto a bath of cast iron increased the temperature, and he saw this as a means of raising the temperature to eliminate carbon to make wrought iron or steel. During his experimental work, he used Swedish cast iron, which was low in phosphorus and sulfur. Later, when his licensees were unable to make steel that was not brittle, Bessemer concluded that the problem was phosphorus and sulfur. Phosphorus could be eliminated by using low-phosphorus iron ore. Robert Mushet, a knowledgeable steelmaker, showed that the addition of manganese was necessary to overcome the cracking during hot working caused by sulfur. The German iron and steelmakers had already developed a master alloy called "spiegeleisen" for adding the manganese. Thus, Mushet was a major contributor to the success of the so-called Bessemer process.

According to Jeanne McHugh in her book on Alexander Holley, Bessemer's American patent was challenged by William Kelly (Ref 1). Kelly filed his own American patent claiming prior development of an air-blowing process. He was able to convince the patent office with documentation and witnesses that his work did precede that of Bessemer. He was awarded a patent with the stipulation that Bessemer had 60 days to contest the decision. Bessemer failed to contest, so

FIG. 3.2 Henry Bessemer, the English inventor who first patented a process and the machinery for blowing air through molten cast iron.

the American patent went to Kelly, but Bessemer retained an American patent on his machinery for blowing air into the molten metal. Kelly received his patent in 1857, just as a financial panic occurred in the United States, and he, along with thousands of other businessmen, went bankrupt.

Kelly moved his experimental operations to the Cambria Iron Company in Johnstown, Pennsylvania, the company where Daniel Morrell and John Fritz were laboring to develop the three-high mill to roll wrought iron into rails. He sold a controlling interest in his patented process in 1861, and two years later a company was formed to produce steel by the pneumatic process (**FIG. 3.3**). This company, the Kelly Pneumatic Process Company, had as principals Captain Eben B. Ward of Michigan, Z.S. Durfee of Massachusetts, Daniel J. Morrell of Cambria Iron Co., and Kelly with a minority interest. Captain Ward was instrumental in building a pilot plant in Wyandotte, outside of Detroit, and William Durfee, cousin to Z.S. Durfee, supervised the operation to produce steel. At this time, America was in a civil war between the industrial north and the agricultural, slaveholding south. Government agents as well as private individuals from both sections of the country were combing Europe looking for armaments and other developments that would aid their cause. One such individual was a charismatic engineer and technical writer by the name of Alexander Holley.

Holley was born in Connecticut in 1832. His father owned a large plant for manufacturing cutlery and was the state's governor in the late 1850s. Holley

FIG. 3.3 Kelly's air-boiling converter for making steel.

graduated from Brown University in 1853 and started work in railroad locomotives. For most of his career, he was a prolific writer of technical articles, mostly for *The New York Times*. He was hired in August 1862 by Edwin Stevens, an armament manufacturer, for a trip to Europe to observe the latest developments in ordnance and armor plate. "The commission from Stevens proved to be the turning point in young Holley's life," as recorded in Jeanne McHugh's book, *Alexander Holley and the Makers of Steel*. On this trip, he came into contact with the Bessemer steelmaking process. Bessemer had a pilot plant where he was very busy in this period showing his process to interested parties. Holley realized that the patent situation had to be resolved before the process could become commercial in the United States.

Upon his return, he contacted two ironmakers through a mutual friend in hopes of interesting them in the Bessemer process. They were John A. Griswold and John F. Winslow of Troy, New York. Troy was a leading iron-producing area with Albany Iron Works, Rensselaer Iron Works, and the Burden Iron Company all located there. They hired Holley to return to England to purchase the rights to produce Bessemer steel in the United States. He was successful in gaining the Bessemer patent rights to the machinery for producing steel, but the patent for the process itself remained with Kelly.

Bessemer Process Enters Production

Holley built an experimental Bessemer steel plant at the Rensselaer Works in 1865. In the meantime, the first successful heat of steel made in the United States by this process was in September 1864 at the little plant of the Kelly Pneumatic Process Company in Wyandotte, Michigan. A plaque honoring this event was awarded by the American Society for Metals in 1976. Ingots from this first heat were shipped to the North Chicago Rolling Mills, where they were successfully rolled into the first steel rails made in America.

The first heat of Bessemer steel was made in February 1865 at the Troy plant just five months after the heat made by Kelly's process. However, the patents held by the Kelly Pneumatic Process Company for commercial production were Kelly's for the process and Mushet's for adding manganese, but not for Bessemer's machinery. The Troy group held patents for Bessemer's machinery, but not for Kelly's process nor for Mushet's manganese addition. This stalemate delayed commercial production by both companies. It was resolved by forming a separate company that held all the patents. The Troy group held 70% ownership and the Kelly group 30%. All royalties were divided on that basis. Release of the patent deadlock cleared the way for development of the U.S. steel industry, and Holley was the chief promoter and designer of this early activity (**FIG. 3.4**).

FIG. 3.4 Alexander Lyman Holley, the developer of Bessemer steel plants in America.

Bessemer Steel Becomes an Industry

Holley then constructed or consulted on nearly all 15 Bessemer plants built by 1875, including the very early plant of the Pennsylvania Steel Company at Steelton near Harrisburg, with J. Edgar Thomson and Tom Scott of the Pennsylvania Railroad Company as principals, and the Cambria Steel Company in Johnstown, Pennsylvania, where Kelly received support to develop his process. Holley built or consulted with plants in North Chicago and Joliet, Illinois, a plant for the Vulcan Works at St. Louis, Missouri, and plants for Bethlehem Steel and Scranton Steel in eastern Pennsylvania. After the building of Bessemer plants in the United States was complete, Holley continued consulting for his clients, exchanging new information about the process and machinery. He traveled throughout the eastern United States and in Europe, where he maintained an interest in developments there.

For his work on the development of the Kelly-Bessemer process, Alexander Holley has been called the "Father of American Steelmaking." He was elected president of the American Institute of Mining and Metallurgical Engineers in 1876 and served as chairman of the first meeting of the American Society of Mechanical Engineers and as vice president of the American Society of Civil Engineers. After his death in 1882 at just 49 years old, these engineering societies commissioned a statue of Holley that remains standing in Manhattan's Washington Square Park (**FIG. 3.5**). Engineers from all of the major technical societies attended the dedication, including members from France and Germany.

These early steel plants were small by later comparison. Bessemer converters, the vessels used to blow air through the steel, had reached only five tons by the early 1870s. Also, their supply of molten metal came mostly from remelting pig iron in separate furnaces. At this stage in America's industrialization, mills were not equipped to handle large quantities of molten metal. Production of pig iron had reached 1,665,000 tons by 1870, but only 68,750 tons were turned into steel. The rest was used as castings or puddled into wrought iron. But the country was poised for much greater things for iron and steel. In the next 30 years, production would increase by a factor of nearly 10 for iron, from 1,665,000 tons in 1870 to 14,000,000 tons in 1900, and an incredible increase for steel, from

FIG. 3.5 Statue of Alexander Holley in Washington Square Park, New York City.

FIG. 3.6 This Bessemer converter at Kelham Island Museum in Sheffield, England, was used until 1974. It is one of only three still in existence.

68,750 tons in 1870 to 10,000,000 tons in 1900 (**FIG. 3.6**). Throughout the 1890s, the United States was producing a third of all the steel in the world, and half of what was produced went into railroad rails. By 1867, steel rails were being rolled commercially at the well-known Cambria Iron Company in Johnstown from Bessemer steel made at the new Pennsylvania Steel Company plant in Steelton. Bessemer steel gradually became the material of choice for rails because it was harder and far more wear resistant than wrought iron. In the final analysis, however, it was the much lower costs for manufacturing Bessemer steel compared with wrought iron that tipped the scales in its favor. The price of steel for rails reached a low of $20 per ton when the competition among the many mills was at its highest. This was the force behind the modern industrial growth of America—low-cost steel. The single individual who was responsible for this drive to lower the cost to produce steel was Andrew Carnegie.

Andrew Carnegie—The Immigrant

Carnegie was born in Dunfermline, Scotland, in 1835 into a family of linen weavers. The 1840s saw the increase of poverty as the demand for the product of hand looms died. This poverty finally drove the Carnegies to abandon their homeland and seek a better life in America. Andrew was 13 years old at the time and had completed only five years of schooling, the full extent of his formal education.

The family arrived in New York City in July 1848, after 50 days at sea. From there, they took a river boat up the Hudson to Albany and then a canal boat for 364 miles on the Erie Canal to Buffalo. A lake steamer took them to Cleveland, and from there they traveled to Pittsburgh, where Mrs. Carnegie had two sisters who had arrived in America some eight years earlier. The Carnegies were penniless and lived on the charity of these sisters.

Both Andrew and his father finally found jobs in a local textile plant at very low pay. Andrew soon obtained another job, which led to clerical work. The new opportunity arose for Andrew as a messenger in the office of the O'Reilly Telegraph Company, earning $10 a month. With his eager ambition, he soon advanced to an operator for which he received $20 a month.

After several years at the telegraph company, Carnegie became acquainted with Tom Scott of the Pennsylvania Railroad Company. Scott was one of those self-made industrial managers for which America became famous during its Industrial Revolution. He came from eastern Pennsylvania and had gone to work for the railroad at the late age of 27. Within two years, Scott was made superintendent of the new Western Division, bringing a single line to Pittsburgh from Philadelphia. Scott was a frequent customer at the telegraph office on railroad business. He decided the railroad should have its own telegraph office in Pittsburgh and offered Carnegie the job of operator. For the next decade, Scott would be a friend and mentor to Andrew Carnegie. Scott was promoted to general superintendent in 1858 and to a vice presidency the next year. When he returned to Philadelphia with this promotion, he recommended Carnegie as superintendent of the Western Division. At age 24, with 11 years in America, Carnegie was now on his own in a management position at the lofty salary of $1500 per year.

It was during these half-dozen years that Carnegie worked directly for Scott that he started to build his financial base. After a modest beginning with a $600 investment in the Adams Express Company, Carnegie next acquired an interest in the Woodruff Sleeping Car Company. There is considerable mystery in this latter investment. Apparently, Scott was instrumental in having the Woodruff stock made available to an unspecified employee of the railroad in return for the railroad buying sleeping cars from Woodruff. Once the contract was signed, the employee turned out to be Carnegie. In addition, the dividend from the stock was sufficient to meet the monthly payments on the loan Carnegie made for the stock purchase. By such techniques, Andrew Carnegie, who never made more than a $2400 annual salary working for the Pennsylvania Railroad, amassed a fortune of $400,000 with annual dividend income of $56,000 by 1867. That year, he and his mother moved to New York City to take up permanent residence in the elegant St. Nicholas Hotel.

Carnegie's early fortune was based on investments in a variety of endeavors. He invested in the early oil boom around Oil City, Pennsylvania, but usually he invested in railroads or companies allied to the railroad activity. Thus, when he saw the superiority of the first iron bridge built in his division of the Pennsylvania Railroad, he induced the key men involved to join him in forming a company to build such bridges. It didn't seem to bother him that they were all employees of the railroad to whom they were selling bridges. Under Carnegie's urging, this investment grew from building small bridges over ravines in the beginning in 1862 to building the famous Eads steel bridge across the Mississippi at St. Louis in 1869–74. The bridge company was reorganized in 1865 as the Keystone Bridge Company with Carnegie's old friend Tom Scott as a subscriber to half the stock.

Carnegie Enters the Iron and Steelmaking Industry

Carnegie's first major investment in ironmaking was simply to supply material to his bridge company. Along with his friend Henry Miller, he founded the Cyclops Iron Company in 1865. Even before it produced any iron, it was merged with a rival in whom Carnegie's brother Tom had an interest. In this merger, Carnegie gained the services of Andrew Kloman, an experienced forge plant operator, and Harry Phipps, an accountant. The merged company was named Union Iron Mills. However, there was a blood feud between Henry Miller and Harry Phipps that Carnegie could not resolve, except by buying out Miller's shares. This gave him controlling interest in the firm, which was renamed Carnegie, Kloman and Company. For a few years, Carnegie's interests revolved mostly around telegraph lines, sleeping cars, bridge building, railroads, and bond selling (**FIG. 3.7**).

During 1872, the first Bessemer plant in Pittsburgh was contracted for by an established ironmaster named William Coleman with his son-in-law Tom Carnegie and several other local businessmen. It was located on a tract of land known as Braddock's Field, where General Braddock had been defeated during the French and Indian War. Andrew Carnegie joined them later "after he traveled to Sheffield to tour Henry Bessemer's steel plant and marveled at his new furnaces," as reported by David Nasaw in his book, *Andrew Carnegie*. He made the largest investment of $250,000 and also tried to interest Tom Scott and J. Edgar Thomson in joining them. However, these men were principals in the Pennsylvania Steel Company in Harrisburg that was already in full production making Bessemer steel rails. Interestingly, Carnegie proposed to name his new steel mill after J. Edgar Thomson, who was probably the most influential railroad executive in America, and he needed the railroads as customers for his new plant. The Edgar Thomson Works was constructed in 1873–75 with Alexander Holley

FIG. 3.7 Andrew Carnegie at the peak of his career as majority owner of the largest U.S. steel producer.

as the engineer in charge. The shop contained all of the improvements resulting from Holley's experience in having already constructed several other Bessemer installations. The Thomson Works was built with two Bessemer converters, each with a five-ton capacity. In addition to the Bessemer converters, Carnegie included two Siemens-Martin open hearth furnaces, also five tons each. These latter furnaces had been developed in Europe about the same time as the Bessemer converter. Unlike the Bessemer process, which could refine a heat of steel in minutes with no added fuel, the open hearth required many hours of processing by an outside heat source. However, over time the open hearth demonstrated advantages that eventually allowed it to replace the Bessemer process. It was surprising to see open hearths in the initial construction of the Thomson Works because they did not become of major importance in the U.S. steel industry for another generation. They were used in the Thomson Works to make small quantities of higher-quality steel when the customer would pay a premium, which was seldom in the early days of steelmaking. If quality was needed, it was still believed that wrought iron was the material of choice. While wrought iron was not as strong as steel, it had the reputation of being more reliable. This was not just a conservative reluctance to try something new; experience in using Bessemer steel was revealing its tendency to fracture on unexpected occasions. It was because of this reputation for reliability that Gustave Eiffel constructed his great tower in Paris of wrought iron as late as 1886.

Captain Billy Jones Joins Carnegie Steel

Andrew Carnegie contracted Alexander Holley as the engineer on the Thomson Works. Holley had been a consultant on the new Bessemer shop at Cambria in Johnstown, Pennsylvania. Holley worked with Captain William (Billy) Jones, who was an assistant to George Fritz, the general superintendent. George Fritz was the brother of John Fritz, who had earlier developed the three-high rolling mill at Cambria. George Fritz died in early 1873, and Billy Jones resigned when he was not promoted to Fritz's job. Billy Jones contacted Holley, who then hired him as his assistant at the Thomson Works. When construction was finished, Carnegie hired Jones as the general superintendent. Jones became another player in a long line of top-quality men Carnegie would attract while forging his great steel empire.

Carnegie's plant construction was just getting underway in 1873 when the most severe economic downturn in U.S. history up to that time occurred. As usual, Carnegie had money in hand to build the mill, but he still needed the bank's help to complete the job. The financial panic of 1873 made it impossible to raise the money needed in this country, so Carnegie and Holley journeyed to London to see their old friends at Morgan and Company. While in Europe, Carnegie's partners back in Pittsburgh found financial backing from local sources. The $400,000 was borrowed based on data furnished by Holley regarding the low cost of making steel at Thomson Works. Profits from the plant repaid the loan within the first year.

During this period when the Bessemer shop was being planned at Braddock, Carnegie was also involved in building a blast furnace for Carnegie, Kloman and Company. This blast furnace was designed on the British practice of one large furnace instead of several smaller ones to save on fuel. It was named Lucy after Tom Carnegie's wife. Lucy stood 75 feet high and 20 feet in diameter and was designed to produce 350 tons of molten metal per week.

A similar furnace was built a short distance away by a group of small manufacturers of cast- and wrought-iron products. This furnace was called the Isabella. The crews of these two furnaces soon engaged in fierce competition in production quantities. This type of worker rivalry is unknown today but was a point of pride in past generations. Week by week, the crews increased production above the design limit of 350 tons. They reached 500 tons in a few months and 600 tons in a year, and by the end of a decade, the two furnaces were producing over 1000 tons per week. This not only greatly increased production but also exhibited the improved blast-furnace engineering technology and maintenance. The blast furnace was forced to keep up with the ravenous appetite of the newly built Bessemer converters.

Carnegie built a second Lucy furnace in the late 1870s and followed with two new furnaces at the Edgar Thomson Works in Braddock a few years later. Nearly all of the increased production went into railroad rails. Carnegie had many contacts within the railroad industry, and he used his influence to promote his product. There was plenty of competition, however, as the other rail producers fought back, but none could meet Carnegie's priority for low-cost production and return of profits. Because he held controlling interest in the various operations of his steelmaking empire, he dictated the use of profits. While other steelmakers paid large salaries and had expense accounts, Carnegie held his top management salaries to $5000 a year and a share in the company. His management team became wealthy in stock, but it paid little or no dividends; the only way they could get their money was to sell to the remaining owners at the original price. Carnegie had them locked in. Eventually he lost many of his good management men and even drove out most of his partners with this business philosophy, including Kloman. Of course, Carnegie and the management team that remained were the final beneficiaries, and ultimately there was plenty of money to share.

The one exception to Carnegie's salary policy was Captain Billy Jones. When the general manager, William Shinn, was forced out of the company for seeking other employment, Shinn tried to take Jones with him. Carnegie offered Jones whatever salary he wanted to stay. Not being greedy, Jones requested $15,000 a year. Carnegie replied by increasing it to $25,000, the same as that of the president of the United States. During that year (1881), the Thomson Works produced over 160,000 tons of rails. Also in 1881, Carnegie reorganized his empire as Carnegie Bros. and Company Ltd. with a value of $5,000,000, and the company made a $2,000,000 profit.

Jones was a truly unique manager. He came from the same mold as John Fritz. He had little formal education and was apprenticed at the very young age of 10, where he learned foundry and machine shop work. He held various jobs in the iron industry and eventually went to work at Cambria Iron Company, where he came in contact with John Fritz and William Kelly. Jones joined the Union Army during the Civil War. His experience and maturity won him a commission, and he was mustered out as a captain at the end of the war.

He returned to Cambria, where he became an assistant to the general superintendent. Jones was a mechanical genius and a true friend to his workers. His patented inventions were a major factor in the great advances of the steel output of Carnegie's empire and of the industry in general during the 1880s. One item alone, the hot car transfer ladle, which moved molten metal from the blast furnace to the Bessemer converters and open hearth furnaces, eliminated the need to remelt pig iron to make steel. Royalties from his many patents earned him $15,000 annually during the late 1880s, and his salary reached $35,000 (**FIG. 3.8**).

His reputation was worldwide. Jones was the first American invited to visit the great Krupp Works in Germany. His greatest legacy, however, was as a leader of men. Undoubtedly his experience leading soldiers was the basic training ground for his concern for his men. Even though he followed Carnegie's commands to get the most possible out of labor, he was the only manager to resist when the demands became excessive. Captain Billy Jones was killed in 1889 at the age of 50 when a furnace exploded. His death was a shock to his fellow workers, the entire steel industry, and the city of Pittsburgh. Steelmaking is said to have ceased while the city mourned at his wake. As reported by McHugh in her book on Alexander Holley, "ten thousand wet-eyed men marched with

FIG. 3.8 Captain William "Billy" Jones, the most popular manager at Carnegie Steel.

him to his grave, and today the veteran steelmaker's most precious memory is 'I worked with Bill Jones.'" Captain Billy lived for his work and for his men.

Carnegie Steel Expands

In the late 1870s, the Thomson Works produced more Bessemer steel than the rail mill could consume. The lack of balance between major processes was a problem in the steel industry. However, at this time Carnegie was able to sell the excess ingots to Pittsburgh area manufacturers who were just switching over from wrought iron to Bessemer steel. As the Braddock plant improved the output of the rail mill so as to consume this steel in-house, Carnegie abruptly cut off his former ingot customers. With nowhere else to turn, they made plans to build a Bessemer plant of their own. As fate would have it, they found Carnegie's embittered ex-partner, Andrew Kloman, to build and manage the plant as part owner. Kloman built a more modern plant than the Thomson Works and also built his own rail mill next door. He was out for revenge. This plant was called the Pittsburgh Bessemer Steel Company and was located in Homestead, just downstream and across the Monongahela from Carnegie's plant.

Andrew Kloman died before the plant was completed and subsequent managers were faced with labor problems that plagued the operation. When the next in a series of recessions hit the rail steel market, the disgruntled owners approached Carnegie to see if he would buy their plant. He was happy to acquire the most modern rail mill in the industry and eliminate a competitor at the same time. In addition, he obtained a mill that rolled structural beams, which became the next product of importance in the coming age of technology. Only later would Carnegie have any regrets over buying a mill that was riddled with labor problems. This new acquisition was called the Homestead Works.

The steadily increasing production of iron and steel in the Monongahela Valley of western Pennsylvania soon exhausted the local sources of iron ore, especially the low-phosphorus type needed for the Bessemer converters. It was not the availability of ore that drove the industry to build in the Pittsburgh area but rather the world's greatest source of coal for making excellent coke. The coal underlies vast regions of western Pennsylvania, West Virginia, and Ohio, but nowhere is the quality of coal for making metallurgical coke better than the coal around Connellsville, just southeast of Pittsburgh. When the era of making iron with charcoal ended, these coal fields with their beehive coke ovens became the major source of fuel in ironmaking. And the major player in this game was a young Pennsylvanian by the name of Henry Clay Frick.

Frick was working as a clerk in his maternal grandfather's distillery (they made the famous "Old Overholt" rye whiskey named after his grandfather, Abraham Overholt), which was located in the heart of the coal fields. He had a cousin who had entered the coal and coke business some years earlier, but the lack of demand before any Bessemer converters had been built, in addition to competition, made it unprofitable. This cousin convinced Frick to join him in saving the business. Frick believed that the operation was too small and that an expansion was necessary for success. He borrowed money from his relatives, especially his wealthy grandfather, to support this questionable venture.

Like so many successful entrepreneurs, Frick was very lucky. Just as he was expanding the manufacture of coke, Carnegie was building his Thomson Works and his Lucy furnace. Carnegie thereby became one of Frick's most important customers. Frick expanded to supply the coke as ironmaking grew in the Pittsburgh area throughout the 1870s. He reorganized in 1881 to raise capital, and Carnegie bought a small interest in the company. However, Carnegie was never one to remain a small shareholder if the business was successful. Within several years, he bought out some of the other shareholders, and the next time Frick needed capital for more expansion he sold even more stock to Carnegie. The controlling interest in the Frick Coke Company thus passed to the Carnegie Bros. Company.

Tom Carnegie, Andrew's younger brother, was managing the company after Shinn was forced out. Tom did not have the instinct for competition that Andrew had, and the heavy responsibility took its toll. Tom became an alcoholic. In late 1886, both Tom and Andrew had serious health problems; Tom had developed pneumonia, and Andrew had contracted typhoid fever. For some time, it was believed that neither would survive. Andrew pulled through, but Tom died. It was after Tom's death that Andrew enticed Frick into the steel company to serve as general manager (**FIG. 3.9**). Thus began the team of Carnegie and Frick that would complete the steel empire that had been slowly growing during the past 15 years.

Once again a group of Pittsburgh manufacturers attempted to build their own Bessemer steel plant in the late 1880s. They were hampered with problems similar to those of the Homestead plant—finances, labor, and Carnegie's devious rumors regarding their poor steel quality. Henry Frick took it upon himself to bargain for this new plant with all of its updated equipment. He finally made a deal for $1,000,000, and the plant was named Duquesne after the town where it was built. Carnegie's empire now contained three major steel mills (Thomson, Homestead, and Duquesne), the Lucy furnaces, the Frick Coke Company, Keystone Bridge Company, and a variety of small mills, mines, and coke works. It was the largest steel operation in the country. Within a few years, by 1892, the output of the newly organized Carnegie Steel Company exceeded half that of all the steel industry of Great Britain. It was a $25,000,000 company with profits averaging 20%. It was the heart of industrial America, and Andrew Carnegie owned over half of it.

FIG. 3.9 Henry Clay Frick, an early coke producer who became general manager of Carnegie Steel.

The Great Strike of 1892 at Homestead

The year 1892 would go down in history for the great strike of the Amalgamated Association of Iron and Steel Workers at the Homestead Works. The entire iron and steel industry awaited the outcome of this settlement because it would set the scales for the other producers. This was the first bargaining experience Frick had with the association, and he was determined to break the union. To ensure control of the situation should force be needed, he made arrangements for 300 guards from the Pinkerton Detective Agency to be available. Frick then rejected the first proposals submitted by the union and offered terms of his own that he knew would be unacceptable. The contract deadline arrived, and Frick started to close the plant to shut out the workers. Both skilled union workers and nonunion unskilled workers walked out.

When the local authorities were intimidated by the laborers, who were joined by the whole town of Homestead, Frick sent for his Pinkerton force. These 300 armed, uniformed guards came down the Monongahela in two steel barges and attempted to land at the Homestead plant. The workers lined the riverbank, and the guards were met by a hail of gunfire. The battle raged all day with four workers killed and many wounded. Late in the afternoon, the guards surrendered, and the union leaders promised them safe passage out of town. However, when they came ashore, they were attacked by members of the union, and as they marched through town the residents unleashed a fury that killed three of the guards and injured many of the rest. Several days later, 8000 state militia troops arrived to take over the plant and return it to management (**FIG. 3.10**). Frick offered to take most of the workers back, but on his own terms. When they still held out, he hired strikebreakers to take their places. Gradually, many of the men rejoined the Homestead Works, but they were forced to work under worse conditions than before. Labor relations and worker morale never recovered in this plant of the Carnegie empire. In an odd twist, a complete outsider to the strike walked into Frick's office a few days later and shot him twice before the gunman was subdued. Frick survived the attack on his life, but his reputation as well as that of Carnegie would never again ride as high within the industrial and political scene.

New Steel Markets Open

The 1890s were a time of rapidly increasing steel production, except for the financial panic that began in 1893. It was a time of change as well. The railroads still were the biggest customers, but their activity had peaked. The railroads had

FIG. 3.10 State militia sent to break the strike at Homestead Works.

FIG. 3.11 Architect and "father of skyscrapers" Louis Sullivan's Prudential Building, Buffalo, New York.

completed the building of major rail lines in the United States, and their needs were now largely for improvements and maintenance. Steel was being tried in many new applications where it permitted breakthroughs in engineering and construction. One such area was the newly developing skyscrapers (**FIG. 3.11**). The 1890s saw such buildings going up in Chicago, the birthplace of these structures, as well as Saint Louis, Buffalo, New York City, and other rapidly growing population centers.

The cost of steel had been driven down by Carnegie and his competitors to the lowest the world would ever see. This was accomplished by every tactic available to an unregulated society, including importing vast numbers of foreign workers to drive wages to bare survival levels. This was the era of the "robber barons." These were the times when companies eliminated as much competition as possible—by crushing them and driving them out of business, buying them out (Carnegie), or creating pools to share orders (Carnegie and the Bessemer producers). But what they all needed was rapid growth in new markets.

And this is what they received because they were standing on the threshold of the American Technological Revolution.

This new revolution had already started in some areas because it filled a need from the older Industrial Revolution. The Bessemer process made steel available for the railroad industry, and the discovery of oil supplied fuel for lighting (kerosene) and industrial lubricants just when whale oil was becoming scarce. And this was the era of building the land grant colleges of the Midwest with their emphases on agriculture and engineering. The groundwork had also already been laid for several brand-new technologies that would revolutionize life in the 20th century.

The first of these new technologies was electricity. It had been studied for over a half century, and laboratory devices had been built in the form of batteries, generators, motors, and measuring devices, and even electroplating was in use as an industrial process. By the 1880s, Edison built the first power plant at Pearl Street in New York City and provided the first commercial electricity for consumer lighting. The availability of electrical power and the electric motor became as revolutionary in the new century as the steam engine had in the previous one.

The invention of the internal combustion engine produced technologies in the 20th century that rivaled the electric age and doomed steam power for almost all applications, except production of electricity. Automobiles, trucks, buses, farm machinery, and airplanes all relied on power based on the burning of oil in the form of gasoline, kerosene, or diesel oil.

All of this, however, was in the future for industrial leaders of the 1880s and 1890s. Their contribution to this coming revolution was to begin building what we now call the "infrastructure" for launching this new age. The steel age was underway, the oil industry with refineries and pipelines was in place, the construction industry was becoming used to longer bridges and higher buildings, and the U.S. Navy was ready to begin a massive program for modern battleships.

Andrew Carnegie Sells His Steel Empire

It was at this critical juncture in American history that Andrew Carnegie decided that he had taken his steel empire about as far as he could take it. He was 65 years old, and his lifelong philosophy had been to turn away from moneymaking for its own sake to do something more useful with his life and fortune. He was also frustrated trying to convince his associates and board of directors to go beyond steelmaking to turn the metal into finished products. During the last few years, he was engaged in an ongoing battle to convince them that future growth lay in producing the very items that his own customers made and for which

FIG. 3.12 Andrew Carnegie in 1913. After selling his company in 1901, Carnegie became a philanthropist and gave away most of the fortune he made building his steel empire.

they bought his steel. Charles Schwab, the new president who replaced Frick, suggested that Carnegie entertain offers for selling his steel empire. Carnegie finally sold to J.P. Morgan of Wall Street fame, who folded Carnegie's company into a new corporation named United States Steel (**FIG. 3.12**).

The Age of Bessemer Steel was beginning to close with the new century. As reported by John L. Bray in his book *Ferrous Production Metallurgy*, "Bessemer steel increased in production between 1900 and 1920 but became a smaller percentage of total production as the open hearth process became the major method of making steel. In the 1920s, Bessemer steel production increased to eight million tons per year while open hearth increased to 40 million tons per year" (Ref 2). Thus, the process that introduced the industrial world to steel now gave way to the new age of making steel in open-hearth furnaces, which would be the process for making steel until the 1960s, when it too would be replaced by another new process, the basic oxygen furnace.

REFERENCES

1. J. McHugh, *Alexander Holley and the Makers of Steel*, The Johns Hopkins University Press, 1980
2. J.L. Bray, *Ferrous Production Metallurgy*, John Wiley & Sons, 1942

Selected References

R.B. Gordon, *American Iron: 1607–1900*, The Johns Hopkins University Press, 1996

T.J. Misa, *A Nation of Steel: The Making of Modern America 1865–1925*, The Johns Hopkins University Press, 1995

D. Nasaw, *Andrew Carnegie*, Penguin Group, 2006

L. Standford, *Meet You in Hell: Andrew Carnegie, Henry Clay Frick, and the Bitter Partnership that Transformed America*, Crown Publishers, 2005

B.C. Stoddard, *Steel: From Mine to Mill, The Metal that Made America*, Zenith Press, 2015

Chapter Four

Metallography—The New Science of Metals

The science of metals, unlike that of chemistry and physics, was slow to develop. Early researchers, especially chemists who were developing ideas and techniques for determining the composition of matter, speculated on the difference between iron and steel. By the late 18th century, Torbern Olof Bergman of Sweden proposed that iron was allotropic (had more than one crystal form) and that steel contained carbon, which made it different from iron. Later, such well-known scientists as the Englishmen Sir Humphry Davy and Michael Faraday devoted some studies to iron and steel, but little understanding came from their work. At about the same time (1820–1830) Dietrich Ludwig Gustav Karsten established that wrought iron, steel, and cast iron differed principally in the amount of carbon they contained, and Justus von Liebig developed a chemical method for measuring the amount of carbon present. Gradually, over the next half century, the role of carbon was more widely understood, and steelmakers began offering steels of controlled carbon analysis. It was the age of the craftsman, however, and the individual skills of the steelmaker determined the success of the product; science had little to offer during this time.

The first technical interests in examining the nature of steel came just past the midpoint of the 19th century. Henry Clifton Sorby of Sheffield, England, examined, polished, and etched surfaces of meteorites and several commercial steels under a microscope during the period 1863–1866 (**FIG. 4.1**). Sorby discovered that the microstructure of steel was complex, and he found an area that he called "pearly" (Ref 1). However, no interest developed at the time even

FIG. 4.1 Henry Clifton Sorby was the first scientist to examine the surfaces of polished and etched steel samples under a microscope.

though he reported his findings in his home town of Sheffield, which had been a toolmaking center for centuries.

Early Studies of the Internal Structure of Metal

The very beginning of any scientific or technical field is always difficult to determine because there is usually some knowledge or activity that can be shown to predate whatever beginning is selected. Thus, while the period 1885 to 1890 can be shown as the beginning of the studies of the internal structures of metals, some previous work had been done by Sorby, Dmitry Chernov, Floris Osmond, Adolf Martens, and others.

It was the first attempts by Osmond and Martens to publish their examination of polished and etched surfaces of steel in the *Journal of the Iron and Steel Institute* of Great Britain that reawakened Sorby's interest in an area he had worked on more than 20 years earlier. Sorby immediately set about a new examination of the microstructure of steels. He presented this work in 1886 and published it in the *Journal of the Iron and Steel Institute* in 1887 (Ref 1). This new work by Sorby, along with that of Osmond and Martens, is considered the real beginning of the field of metallography, the study of the internal structure of metals by light-reflected microscopy. From this point forward, the ever-increasing research in how the behavior of metals relates to their internal structure has been the foundation of our modern technological age.

Shortly thereafter, a new technique for measuring high temperatures was developed by Henri Le Chatelier of France with the use of platinum-platinum/rhodium thermocouples. Osmond, who worked with Le Chatelier for many years at the Sorbonne, immediately put the new thermocouple to use in measuring the so-called critical temperatures in steel. The Russian metallurgist D.K. Tchernoff first noted that there were changes in the rates of cooling or heating. He stated that steel could not be hardened upon quenching until it was first heated above the uppermost critical temperature. These critical temperatures were believed to represent important internal changes in steel, but the nature of these changes was beyond the ability of these early pioneers to decipher; the science of metallurgy was too new. In fact, the final solution to these questions about internal changes during heat treatment lay more than 40 years in the future and a continent away. Still, these first few metal makers, with their intense desire to understand the mystery of metallurgy, devoted the rest of their working careers to the endeavor and slowly added to the knowledge base.

In all of metal making, nothing exceeds the technical importance, scientific complexity, and human curiosity surrounding the hardening of steel. The magnitude of strength properties in hardened alloy steel can be seen when compared to plain carbon structural steel. Structural steel has a yield strength (the point on a stress-strain curve where plastic deformation starts) of about 40,000 pounds per square inch (psi) of cross-section; heat-treated alloy steel has a yield strength of 150,000–200,000 psi, increasing to as high as 250,000–300,000 psi for special applications and to 400,000 psi and higher for tools and bearings. Hardened alloy steel is thus a metal of enormous versatility.

Meanwhile, a discovery beyond the field of metallurgy occurred during these early times. An American research professor named Josiah Willard Gibbs (**FIG. 4.2**), at Yale University, developed a theory called "the equilibrium of het erogeneous substances." It received very little notice in the field of science and none in metallurgy.

However, a generation later, Dutch metallurgist H.W. Bakhuis Roozeboom applied Gibbs's theory to develop the iron/iron carbide (Fe/Fe_3C) phase diagram (**FIG. 4.3**). This was the first quantitative experiment to explain what the metallurgists saw in their microscopes. The polished and etched surfaces showed a white phase that was nearly pure iron. With the addition of carbon, a new phase occurred that Sorby had called "pearly." Osmond named this phase pearlite in honor of Sorby.

Both Gibbs's theory and metallography showed that pearlite was a mixture of iron and iron carbide and that the higher the carbon content, the more

FIG. 4.2 Josiah Willard Gibbs developed the theory for systems that were not mutually soluble, that is, that had more than one phase present at equilibrium.

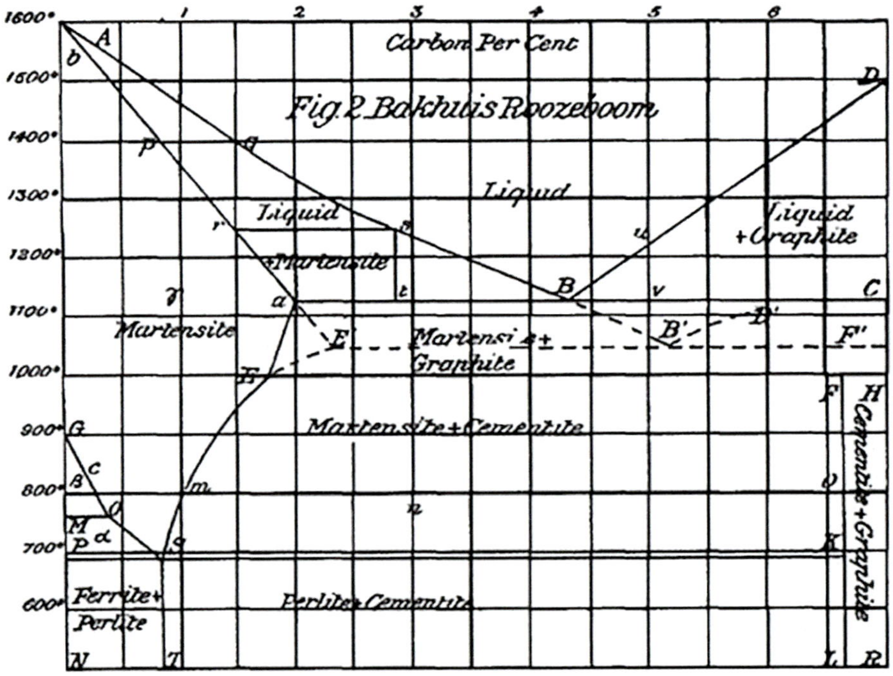

FIG. 4.3 The iron/iron carbide phase diagram as produced by H.W. Bakhuis Roozeboom.

pearlite appeared in the microstructure. At roughly 0.80% carbon, the steel microstructure became all pearlite. Roozeboom's phase diagram showed that the upper critical temperature, reported by Tchernoff, depended on the amount of carbon. The higher the carbon content, the lower the upper critical temperature until the temperatures joined at the lower critical point of 690 °C (1275 °F).

From the iron/iron carbide phase diagram, the transformation of steels of various carbon contents followed. For instance, a 0.20% carbon steel starts to form ferrite (named from the Latin word for iron) from austenite at 835 °C (1535 °F). As the temperature slowly decreases, additional ferrite forms. Because the ferrite is essentially free of carbon, the excess carbon diffuses into the remaining austenite. When the temperature reaches the lower critical 690 °C (1275 °F), the maximum amount of ferrite has formed, and the remaining austenite contains 0.80% carbon, named the *eutectoid composition*. Eutectoid comes from the Greek term meaning transformation from a single phase (austenite) into two phases (ferrite plus iron carbide) simultaneously. This explains the appearance of pearlite with the parallel layers of ferrite and iron carbide (**FIG. 4.4**).

The phase diagram shows phases present under very slow cooling, approaching equilibrium conditions. Therefore, it does not show a phase for martensite because martensite occurs with very rapid cooling (**FIG. 4.5**). Roozeboom named the high-temperature phase martensite (for the German metallurgist Adolph Martens) because it was believed at that time that the high-temperature microstructure was retained on rapid quenching. That was shortly

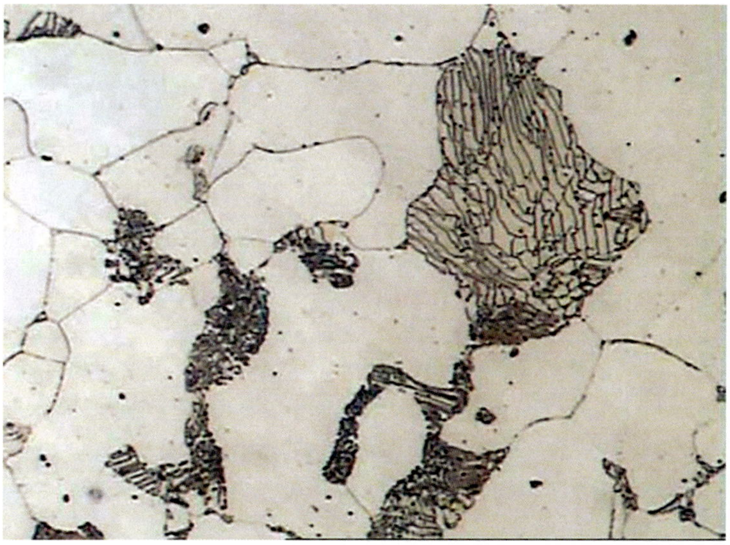

FIG. 4.4 Microstructure of 0.20% carbon steel. The white phase is ferrite, the phase with alternate plates of white and dark is pearlite, and the dark lines are boundaries between crystals of different orientations.

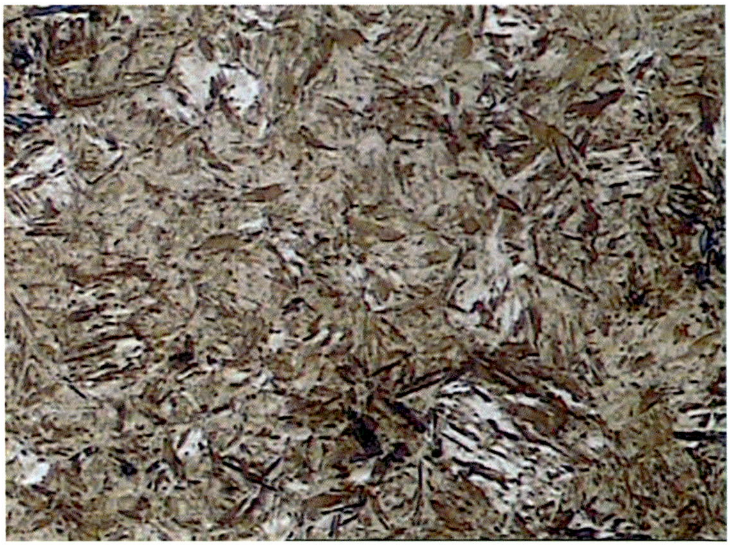

FIG. 4.5 Martensite microstructure. The transformation of austenite to martensite was not understood until much additional research was performed.

corrected, and Osmond named this high-temperature phase austenite, after the English metallurgist Sir William Chandler Roberts-Austen. Roozeboom's work started a trend of measuring the phase relationships in many alloy systems. Much of this work was performed in Germany and England.

X-Ray Diffraction

One of the major achievements of the scientific world that had a large impact on the field of metals occurred just prior to World War I. This was the discovery that x-rays could be used to measure the structure of crystalline solids. The concept was very simple yet profound. Ever since the discovery of x-rays, scientists had theorized about their nature. Eventually, it was proposed that x-rays exhibit some properties of waves, such as light, but with wavelengths. Wavelengths are the distance from any point on the wave curve to the same point on the next cycle. German physicist Max von Laue proposed that perhaps the wavelength of x-rays could be diffracted and reflected by the planes of atoms making up the structure of crystalline solids just as manmade diffraction gratings were used with light waves. This was a scientific proposal proven by Laue when he aimed a beam of x-rays at thin slices of crystalline material. Photographic film mounted behind his samples picked up the main beam coming through the sample, but it also recorded spots from diffracted radiation.

This diffraction phenomenon was reduced to a useful quantitative tool by the father and son team of English physicists W.H. and W.L. Bragg (**FIG. 4.6**). For

FIG. 4.6 William Henry Bragg (pictured) and his son developed the method of using x-rays to measure the distances of atoms in solid metals.

their work on the mathematical solution to x-ray diffraction of crystalline solids, the Braggs received the Nobel Prize in Physics in 1915.

Albert W. Hull in the United States and Peter Debye and Paul Scherrer in Germany developed a technique of passing x-rays through powder samples, which became a common method of studying metals.

At last, the metallurgists had a powerful new tool for determining the internal structure of metals. Among the earliest Americans to apply x-ray diffraction to the study of metals was Edgar C. Bain. He showed that steel heated to the hardening temperature had a different crystal structure. See chapter 5, "The History of Engineering Alloy Steels," in this book for more information and (Ref 2).

The Discovery of Precipitation Hardening

Precipitation hardening was discovered in the early 1900s by a German engineer named Alfred Wilm (**FIG. 4.7**). He was born in 1869 in Nieder-Schellendorf,

FIG. 4.7 Alfred Wilm discovered a new method of hardening metals.

Germany. Wilm was interested in chemistry at an early age and later attended the Royal Institute of Technology at Charlottenburg, where he was introduced to metallurgy. In 1901, he joined the government Central Laboratories for Scientific Research at Neubabelsberg near Berlin.

Wilm began a study to replace the heavy brass alloy, the time-honored jacket material for ammunition cartridges, with an aluminum alloy for weight savings. None of the aluminum alloys up to this time were strong enough for the application. Wilm, who was familiar with heating and quenching steel, attempted to combine both alloying and heat treating in his research. No heat treatment known at that time could harden metals besides steel. Wilm was working with aluminum-copper alloy heating samples, heating them to a high temperature and water quenching them to room temperature. The strength of his heat-treated samples was too low to interest the army. To increase hardness, he tried adding one-half percent magnesium to his alloy. After heating and water quenching, Wilm reported that he gave a sample to his assistant, Fritz Jablonski, for hardness testing. Because it was a Saturday afternoon, Jablonski wanted to leave the laboratory early. Wilm urged him to take at least one reading and then finish on Monday. The reading on Friday showed only a small increase in hardness due to the magnesium addition. On Monday, however, the hardness was considerably higher.

After Wilm and Jablonski double-checked the results, they repeated the experiment, taking hardness readings on the heat-treated alloy every hour. Wilm reported, "For almost two hours I saw no change, but after that time the hardness began to increase, which I followed from morning to evening. The next morning I continued, and determined that after about four days the material was quiescent" (Ref 3). Wilm and Jablonski had observed a room-temperature hardening (or strengthening) with time in a metal alloy, something never seen before by metal makers. Wilm's heat treatment was called age hardening (**FIG. 4.8**).

Interest in Wilm's alloy, called duralumin, spread throughout the metal making world. Germany built 100 zeppelins during World War I made of duralumin, which were used to bomb London. Quantities were used in England and France during the war for aircraft parts, but major airframe structures were still made of wood and canvas. In 1918, studies were undertaken on the hardening mechanism at the United States Bureau of Standards in Washington. Several young metals researchers working at the bureau at the time were Paul Merica, Howard Scott, and R.G. Waltenberg. These men would become famous as the next major contributors to the study of age hardening.

The Research Program

Their study was designed to examine the role of copper and magnesium in aluminum independently and then combine them in various amounts to

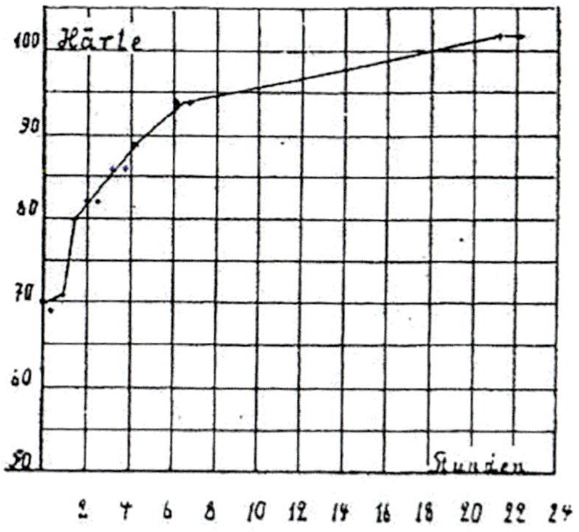

FIG. 4.8 The first age-hardening curve determined by Alfred Wilm in 1899. Source: Ref 3.

determine how they interacted to cause age hardening. Samples were aged at room temperature for 11 days and at 100 °C (212 °F) for three days. The alloys with up to 3.5% Mg and no Cu showed no age hardening even though there was a precipitation of MgAl compound after long aging times at the higher aging temperatures. The Cu-containing alloys with up to 3.2% Cu had an increase in ultimate tensile strength (the stress where failure begins) from 22,000 psi as quenched to 30,000/35,000 psi after aging at both room temperature for 11 days and 100° C after three days. After longer aging times at higher temperatures, a precipitate of $CuAl_2$ was visible in the microscope, and the mechanical properties decreased. Copper and magnesium in combination aged to ultimate strengths of 40,000–60,000 psi depending on the various amounts present in the alloy. The higher strengths occurred after aging at 100 °C with 3–4% Cu and 1–1.25% Mg.

Metallographic studies showed no evidence of a precipitate for samples aged at room temperature and for samples aged at 100 °C up to the maximum hardness. Beyond that time at 100 °C, they found a precipitate of $CuAl_2$ that grew in particle size with longer aging times. They finally turned to solubility studies, which showed that at 500 °C (932 °F), the normal heat-treating temperature for quenching the 4% Cu, the alloy was a solid solution with the Cu atoms substituting for Al atoms in the crystal lattice. But the solubility decreased to 2% at 300 °C (572 °F); extrapolating the solubility line to 100 °C, the solubility was only 1%. The quenched alloy of 4% Cu held 3% Cu in supersaturated solution, which was unstable at room temperature and 100 °C. Merica and his team concluded that this supersaturated solid solution formed a pre-precipitate that hindered plastic deformation until higher stress was applied to the material. The hardness decreased when overaging at higher temperatures where $CuAl_2$ could be seen.

The authors stated, "If the temperature of the alloy is raised to 100 °C, or even to ordinary room temperature, according to the theory which the authors propose, the mobility of the molecules becomes sufficiently great that precipitation of the $CuAl_2$ takes place in the form of very fine particles of colloidal dispersion" (Ref 4). The authors finally concluded, "Upon aging a quenched sample at 200 °C, the hardness first increases to a maximum and afterwards decreases. During that aging there has been first a formation of fine nuclei of $CuAl_2$ followed by coalescence of these particles into ones of larger size. There is, therefore, a certain average size of particle of $CuAl_2$ for which the hardness is a maximum; atomic dispersion of the solute is not the dispersion that produces the maximum hardness but some intermediate one between it and that at which the particles become visible by ordinary means. To this dispersion is due the hardening of duralumin" (Ref 4).

The results of this study by Paul Dyer Merica and his co-researchers were quoted by Zay Jeffries as "the shot heard around the world." If duralumin could be hardened by precipitation of a compound, then this new hardening mechanism might be applied to many other alloy systems. The resulting developments opened the way to a whole new world of strong alloys.

Research on the Mechanism of Age Hardening

A better understanding of metal behavior during plastic deformation began in the late 1920s and 1930s with the research of such Europeans as Geoffrey Taylor, Alan Griffith, M. Polanyi, U. Dehlinger, and E. Orowan. These metals scientists developed the concept of a crystal lattice defect they called a dislocation. Dislocations are segments of a row of atoms that do not line up with the atoms on the next row as they would in a perfect crystal. These dislocations move if enough stress is applied to the metal. The movement or gliding of dislocations accounts for plastic deformation in metals. Dislocations interact with one another and with grain boundaries; they multiply by various means, which explains the shape of the plastic deformation curves.

The concept of dislocations as the instrument of plastic deformation was also useful in explaining the strengthening of alloys by precipitation hardening. If dislocations moving under an applied stress produce plastic behavior, then anything restricting this movement increases the yield strength. The method by which precipitates accomplished this result became the challenge for many metals researchers from 1930 to the 1960s.

Starting in the mid-1930s, there were several successful studies that identified the initial development of precipitation. Researchers began to see anomalies in x-ray diffraction studies that preceded the formation of the final precipitate. Two researchers in Europe, A. Guinier and G.D. Preston, independently identified the first stage of the movement of atoms to sites in the lattice in preparation for forming the final precipitate. These sites were rich in copper atoms in an aluminum-copper alloy but still retained the crystal structure of aluminum. The sites of this first stage in precipitation were called G-P zones in honor of Guinier and Preston. As these zones grew, the strength continued to increase as long as coherency was maintained with the aluminum fcc crystal lattice. When they grew larger with additional aging and formed a nonmetallic compound with their own boundary between them and the aluminum, the strength decreased. These are larger particles, and the distance between them allowed dislocations to move more freely through the metal.

Metals researchers now had a number of tools to study the structure of metals and how that structure related to physical and mechanical properties. The light microscope, phase diagram, x-ray diffraction, principles of precipitation hardening, and dislocation theory would all be used in the coming years to understand and tailor new alloys of high strength for modern technology.

REFERENCES

1. H.C. Sorby, On the Microscopical Structure of Iron and Steel, *J. Iron and Steel Instit.*, (No. 1), 1887, p 255–88
2. E.C. Bain, Metal Structure and the Hardenability of Steel, 1920–1940, *The Sorby Centennial Symposium on the History of Metallurgy*, C.S. Smith, Ed., Gordon and Breach Science Publishers, 1963
3. A. Wilm, Physico-Metallurgical Investigations of Aluminium Alloys Containing Magnesium, *Metallurgie*, Vol 8, 1911, p 225–227
4. P.D. Merica and R.G. Waltenberg, *Heat Treatment of Duralumin*, The Science Paper, US Bureau of Standards, Washington, DC, 1919

Selected References

A. Guinier, Structure of Age-Hardened Aluminum-Copper Alloys, *Nat.*, Vol 142, 1938, p 569–70

H.M. Howe, *Metallurgical Laboratory Notes*, Boston Testing Laboratories, 1902

Z. Jeffries, Autobiographical Notes of a Metallurgist, *The Sorby Centennial Symposium on the History of Metallurgy*, C.S. Smith, Ed., Gordon and Breach Science Publishers, 1963

Z. Jeffries and R.S. Archer, *The Science of Metals*, McGraw-Hill Book Company, 1924

P.D. Merica, R.G. Waltenberg, and J.R. Freeman, *Constitution and Metallography of Aluminum and Its Light Alloys with Copper and with Magnesium*, The Science Paper, US Bureau of Standards, Washington, DC, 1919

W.D. Mogerman, Zay Jeffries, American Society for Metals, 1973

E. Orowan and K.J. Pascoe, An X-Ray Criterion for Distinguishing Lattice Curvature and Fragmentation, *Nat.*, Vol 148, 1941, p 467–70

G.V. Parr, The Criteria of the Martensite Transformation, *The Sorby Centennial Symposium on the History of Metallurgy*, C.S. Smith, Ed., Gordon and Breach Science Publishers, 1963

G.D. Preston, Structure of Age-Hardened Aluminum-Copper Alloys, *Nat.*, Vol 142, 1938, p 570

V.D. Sarhovsky, Tschernoff and the Development of the Theory of the Heat Treatment of Steel, *The Sorby Centennial Symposium on the History of Metallurgy*, C.S. Smith, Ed., Gordon and Breach Science Publishers, 1963

C.S. Smith, Ed., *The Sorby Centennial Symposium on the History of Metallurgy*, Gordon and Breach Science Publishers, 1963

H.C. Sorby, On the Microscopical Structure of Iron and Steel, *J. Iron and Steel Instit.*, (No. 1), 1887, p 255–88

G.I. Taylor, Mechanism of Plastic Deformation of Crystals, *Proc. R. Soc. London* Vol 145 (No. 855), 1934, p 362–87

Chapter Five

The History of Engineering Alloy Steels

Alloy steels were first developed during the late 19th and early 20th centuries. They did not debut until steel itself was readily available, and that did not take place until after 1870. The only alloy steel before this time was a tool steel invented by Robert F. Mushet in 1868. Mushet's steel was very high in both carbon (2% or more) and alloy (7% tungsten and 2% manganese) and was considered a unique material, not a forerunner of modern-day-engineering alloy steels.

However, the idea of alloying was old. It had been practiced since ancient times with copper and tin to make bronze, with gold and silver, and with copper and many other metals. Most mineral sources available contained combinations of metals so that alloys were the natural output of many of these ancient metalworking activities. The carbon in cast iron and steel automatically makes them alloys. Wrought iron is the only nonalloyed iron. Carbon is widely considered to be the single most important alloying element added to iron. Amounts as small as 0.05% have profound effects on the behavior of iron, and 0.10 to 0.25% additions are sufficient to make mild steel, the common steel used in construction. Most heat-treated alloy steels contain 0.30 to 0.40% carbon, while tool steels contain 0.60 to 1%.

Carbon has two characteristics that account for the powerful effects of such small amounts. It is very low in density, so there are a great number of atoms in these small amounts of alloy additions by weight. Further, the atoms are small in size compared to iron so they do not substitute for an iron atom in the crystal lattice but take up a unique position in the holes between the iron atoms. It is this interstitial position of carbon atoms in the iron lattice, along with the crystal lattice transformation that occurs upon heat treating from face-centered-cubic (fcc) to body-centered-cubic (bcc), that makes steel a unique material for construction, power transmission, and tools in modern technology.

Chromium Alloy Steels

The first engineering alloy steel was patented in 1865: Patent No. 49495 was filed by an American named Julius Baur and manufactured by the Chrome Steel Company of Brooklyn, New York (Ref 1). Captain James B. Eads, the builder of iron-clad boats for use on the Mississippi River during the Civil War, selected this chromium alloy steel for use as the major structural material in his famous bridge across the Mississippi in St. Louis, Missouri, in 1869. This was a daring and costly decision for Eads. It was daring because no one had used steel for such a large structure, let alone used the totally unknown alloy addition of chromium, and costly because Eads used crucible steel, a process primarily used for making tool steels. Crucible steel cost about $300 per ton compared to $80 for wrought iron. The Eads Bridge has stood the test of time for over 145 years.

The publicity surrounding the Eads Bridge and the use of chromium alloy steel prompted a French metallurgist, Henri-Ami Brustlein, to become interested in making chromium alloy steels. Brustlein soon learned that to alloy chromium with steel he first needed to refine the chromium ore to produce what is called ferrochromium, a master alloy of iron-chromium-carbon. This master alloy would readily dissolve into the melt during the crucible process; this was important because otherwise the recovery of chromium would be too erratic to control the alloy content. Brustlein produced and sold chromium alloy steels for tools, cannon shells, and armor plate over a period of 15 to 20 years before anyone else. In addition, he replaced the manganese with chromium in Robert Mushet's special tool steel to improve it. He brought samples of the new tool steel to the United States where he interested Fredrick Taylor at Midvale Steel Company in its use. Taylor went on to develop the first high-speed steel, which revolutionized the machining industry. Thus, Brustlein laid the groundwork for both engineering alloy steels and modern tool steels.

Nickel Alloy Steels

During the period that Brustlein was developing chromium steels, other French metallurgists were learning to refine nickel-containing ore from New Caledonia, a French territory in the South Pacific. The nickel was then used as an addition to steel. The production of nickel steel was observed in France in 1888 by James Riley, an Englishman who made arrangements for similar steels to be made at the Steel Company of Scotland in 1889. He immediately tested these steels and reported their properties in the *Journal of the Iron and Steel Institute*. One of the steels containing 0.2% carbon and 5% nickel developed strength properties of considerable interest for many different structural and machine applications.

This steel, processed by slow cooling after rolling, was about 40% stronger than similar steel without nickel. Almost immediately, nickel steels became of interest for armor plate in France (5% nickel) and in Germany (7% nickel). The United States Navy arranged a famous shootout in 1890 at Annapolis where eight-inch armor-piercing shells were fired at close range to compare carbon steel, nickel steel, and a special British compound-carbon steel. Nickel steel was the clear winner. Shortly thereafter, Bethlehem Iron Company produced nickel steel in forged armor plate and other forged parts including the 45-ton shaft for George Ferris's famous wheel, which he built for the Chicago World's Fair in 1893 (**FIG. 5.1**). This first Ferris wheel was not the small carnival type we are used to seeing but a giant machine carrying 60 people in each of 36 cars.

FIG. 5.1 The Ferris wheel at the 1893 Chicago World's Fair.

Nickel steels were used that same year for shafts on several German-built ships for the North Atlantic.

The first alloy steel used in regular industrial production in the United States was a 5% nickel steel for bicycle chain (1898) by the Whitney Manufacturing Company of Hartford, Connecticut, followed the next year by bicycle tubing by the Pope Manufacturing Company (**FIG. 5.2**). Somewhat later, nickel steels (3.5%) became popular for the structural members of large bridges, including the Manhattan and Queensboro bridges in New York City.

Nickel-Chromium Alloy Steels

Shortly after nickel steels came into use, more complex alloy steels containing both chromium and nickel were being used by the Krupp Steel Works in Germany and by the Compagnie des Forges de la Marine in France. These nickel-chromium steels could be hardened in large sections by heat treating so they became very popular for armor and large forgings. After the turn of the century, the straight nickel steels rapidly declined in use in favor of the nickel-chromium steels and the newly developed chromium-vanadium steels.

FIG. 5.2 The Pope bicycle used 5% nickel alloy steel for the frame.

Automotive Demand for Alloy Steels

It was the automobile, however, that introduced the age of alloy steels. The Haynes Automobile Company (formerly known as the Haynes-Apperson Company) took the lead in specifying a 5% nickel steel for axles (**FIG. 5.3**). The first attempt at alloy steel specifications was by a group known as the Association of Licensed Automobile Manufacturers. In 1910, they issued specifications written by Henry Souther for a nickel steel, two nickel-chromium steels, and three carbon steels. Elwood Haynes had reported earlier in the American Society of Mechanical Engineers (ASME) journal *Transactions* (1907), on using a nickel-chromium steel for gears. He said, "Gears were made of the best kind of tool steel without success because the ends of the teeth would break off. However, gears made from nickel-chromium steel, if properly heat treated, would run for an entire season without the breakage or serious injury of a single tooth" (Ref 2). Early motorists were accustomed to the annual replacement of gears, axles, bearings, and other highly stressed parts.

FIG. 5.3 The Haynes-Apperson Company used 5% nickel alloy steel for the first time in an automobile. Source: Brian Snelson.

Development of Specifications for Alloy Steels

By 1912, both the new Society of Automotive Engineers (SAE) and the American Society for Testing and Materials International (ASTM) developed standard specifications for alloys containing nickel, nickel-chromium, and nickel-chromium-vanadium. These two societies continued to develop alloy steel specifications, not only for the automotive industry but for industry in general. These specifications were usually written by committees assembled by the societies representing individuals from both suppliers and users of the various alloys. Alloy steel specifications are never just single compositions in chemistry but represent a range for each element, which is a compromise between what the user would like to have and what is practical for the producer to make (**TABLE 5.1**). They also put limits on carbon, silicon, phosphorus, sulfur, and sometimes other elements that are added or considered impurities.

A typical specification for an alloy steel is shown in Table 5.1. This specification states that each element can vary within the range shown and that the impurities, sulfur and phosphorus, must be no more than 0.40% each. This alloy can be ordered anywhere in the world to the SAE/ASTM specification and will be furnished to this composition range. Tables of all the carbon steels and the engineering alloy steels under the SAE/ASTM system are published and used by all of American industry and many foreign countries. The number of alloy steels used in the auto industry increased slowly throughout 1910–1920. Walter Jominy, a metallurgist from the University of Michigan who worked for the Studebaker Car Company, published a list in 1920 of 12 alloy steels that he said filled all the needs for building automobiles.

The use of alloy steel increased from 3000 tons of mostly nickel steel in 1900 to about 150,000 tons by 1910. Of this larger quantity, 40% was nickel steel, about 30% was nickel-chromium steel, and roughly 25% was chromium-vanadium steel, which was introduced in 1905. Later, the addition of 2–3% molybdenum was found to suppress an embrittlement that occurred during the heat-treating operation.

TABLE 5.1 Specifications for AISI/SAE 4340 alloy steel

Chemical composition, wt. %						
C	Ni	Cr	Mo	Mn	Si	P and S
0.38–0.43	1.65–2.00	0.70–0.90	0.20–0.30	0.60–0.80	0.20–0.30	0.40 max

C, carbon; Ni, nickel; Cr, chromium; Mo, molybdenum; Mn, manganese; Si, silicon; P, phosphorus; S, sulfur

World War I placed additional importance on the use of alloy steels and on the process of heat treatment. The quantity of alloy steel made in the United States reached over one million tons in 1918. The 1920s saw a continuing increase in the use of alloy steels, mainly due to the increasing production of autos, trucks, tractors, and other consumer goods. By 1929, auto production reached five million cars, a large number even in the first years of the 21st century. While the cars and other vehicles of the 1930s seem simple to us now, they were becoming much more durable and reliable. Many of the improvements were due to the greater use of heat-treated alloy steel in suspension systems, steering, power transmission, and engine components. This, along with general improvement in engineering designs, a fully enclosed body (made possible by the continuous rolling mill), improvements in hydraulic shock absorbers and brakes, and the introduction of six-cylinder engines, enabled production of a vehicle that is the direct ancestor of modern automobiles.

An Organization for Heat Treaters

The use of alloy steels brought about a significant need for the process of heat treating. The term *heat treating* has many meanings, but the principal one for alloy steels is the heating to a high temperature to form the fcc phase called austenite, rapidly quenching to ambient temperature to transform to martensite, which is the microstructure of hard steel, and reheating (tempering) to produce a set of properties suitable for the intended application. This treatment for steel has been used for hundreds of years and practiced as an art. The art was closely guarded and generally passed down from father to son from generation to generation. As could be expected, results varied considerably. Control of variables such as carbon content, heating temperature, quenching media, and tempering temperature was beyond the technology of the times.

The growing use of alloy steel increased the need to acquire and share heat-treating information. William Park Woodside, a former blacksmith, began to hold meetings in Detroit to exchange information among heat treaters in the early auto industry. These meetings led to the formation of a formal group called the Steel Treaters Club, which later was called the American Society for Steel Treating, the American Society for Metals, and eventually ASM International. The early groups formed chapters in various industrial regions across the country and published data sheets on the technical aspects of heat treating. They soon issued a new publication, *Transactions*, for serious researchers to publish their papers. In 1930, the American Society for Steel Treating published a magazine called *Metal Progress* that became the most popular source of information in all of metalworking, not just heat treating. ASM International is located in Cleveland and continues to publish a vast amount of information on all aspects of metallurgy, organize conventions and symposia, and conduct training classes for metallurgists and nonmetallurgists on the technology of metal processing.

The theory of the hardening mechanism of heat-treated steel saw little progress during the first two decades of the 20th century. While this lack of scientific advancement did not deter the extremely rapid use of heat-treated alloy steel, it led to many alloys with emphasis on special alloy compositions providing superior properties. No two manufacturing companies seemed to use the same alloy steel for the same application, a wasteful use of expensive alloying elements.

Research on the Hardening of Steel

The first research published that brought the older ideas into question was a paper in 1919 by a well-known French metallurgist, Albert Portevin, and his coauthor M. Garvin (**FIG. 5.4**). They showed for the first time that the transformation to hard martensite did not occur until the steel being quenched had cooled

FIG. 5.4 Albert Portevin, the French metals researcher who published the first study on the modern theory of the hardening of steel.

FIG. 5.5 Edgar C. Bain was the first to study the crystal structure of steel by x-ray diffraction.

to temperatures well below that where the two-phase microstructure of pearlite formed. A few years later (1922), American metallurgist W.R. Chapin showed that a carbon tool steel could be quenched to 300 °C (570 °F) and still be austenitic; further, with slow cooling below 570 °F, it could be studied as it gradually transformed to martensite upon falling temperature (Ref 3). These two early studies were of special interest to Zay Jeffries, Edgar Bain, and Marcus Grossman, who were studying the transformation of austenite to lower temperature phases.

The first research to define the crystal structure of the different phases in steel was by Edgar C. Bain (**FIG. 5.5**). Using x-ray diffraction on samples heated to the high temperature phase (austenite), he found they were face-centered-cubic (fcc). Ferrite and martensite at room temperature were body-centered-cubic (bcc). This explained the higher solubility for carbon in austenite compared with ferrite. The interstitial holes in austenite are much larger than those in ferrite.

The next major scientific advance in hardened steel was the discovery through precision x ray diffraction by W.L. Fink and E.D. Campbell that showed that martensite was not a simple bcc structure as reported by Bain but a distorted structure called tetragonal. Professor E.D. Campbell of the University of Michigan was a chemist-turned-metallographer who was blinded early in his career by a laboratory accident. The prestigious annual Campbell Memorial Lecture was named in his honor by ASM. At last, after 35 years of studying martensite, there was now some rationale for its great hardness and lack of ductility. Unlike ferrite, which has very low solubility for carbon, or pearlite, which has layers of iron and iron carbide, martensite traps the carbon on an atomic scale within its crystal structure.

The Transformation of Austenite

The next year (1930), Edgar C. Bain and E.S. Davenport of the U.S. Steel Corp. Research Laboratory at Kearny, New Jersey, published their famous paper, "The Transformation of Austenite at Constant Subcritical Temperature" (Ref 4). What Bain and Davenport did that was totally new was to quench a series of thin samples from the hardening temperature to a transformation temperature and hold them at this temperature for various times before water quenching to martensite. By using metallography and thermal expansion measurements, they were able to follow the progress of formation of the new transformation product from the very beginning, throughout the reaction, and to the end. They plotted individual transformation curves as a function of time at each transformation temperature.

At last, the sequence of events occurring during transformation could be watched, accurately described, and quantitatively measured as the austenite transformed to a variety of structures depending on the transformation temperature. The diagram resulting from plotting the beginning and end of transformation at various temperatures was called a time-temperature-transformation diagram, or TTT diagram; many of these diagrams were determined for numerous alloy steels in the years following the original work of Bain and Davenport (**FIG. 5.6**). The TTT diagram is not the most useful tool for practical heat treating, but it satisfies the deeper desires of the steel metallurgist for a more basic understanding of the transformation of austenite.

Bain's studies showed that austenite transformed to ferrite and/or coarse pearlite at higher temperatures of 650–705 °C (1200–1300 °F) and a finer pearlite at intermediate temperatures of 480–540 °C (900–1000 °F). At still lower temperatures, a unique structure formed that was first reported by J.M. Robertson in the year before Bain's work was published. Later, Bain's fellow workers at U.S. Steel Corp. honored him by calling this new structure *bainite*. This nomenclature was officially adopted. Edgar C. Bain, America's most outstanding metallurgist, was the only native whose name has passed into the daily nomenclature of the metal maker.

The transformation of austenite to pearlite was a subject of intensive study during the late 1930s at the Carnegie Institute of Technology (now Carnegie Mellon University) in Pittsburgh under the direction of Professor Robert F. Mehl. A group of graduate students under Mehl published a series of papers on the mechanisms and rates of formation of pearlite nodules. This work culminated in a mathematical solution developed by W.A. Johnson and R.F. Mehl that could describe the pearlite reaction. Later, Mehl would be the first American to use the electron microscope for metallography.

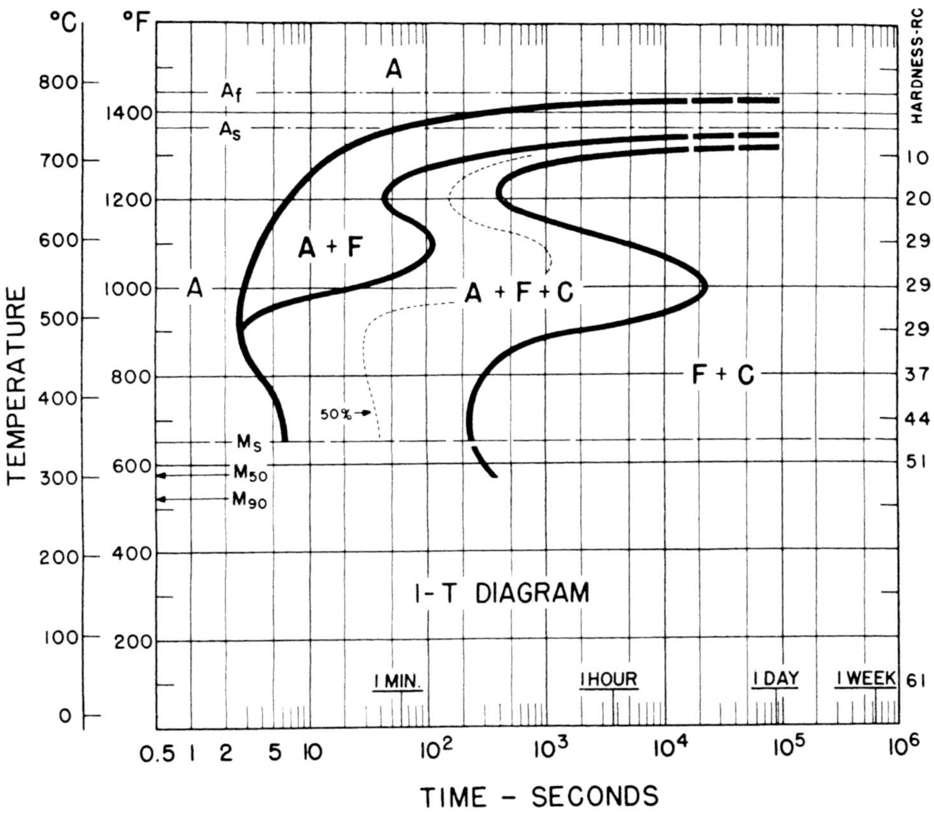

FIG. 5.6 TTT Diagram for 0.80% carbon steel transforming to pearlite.

The Solution to Hardenability Testing

Throughout the 1920s and 1930s, other American metallurgical investigators were working on the more practical aspects of heat treating steel. Howard Scott and H.J. French at the National Bureau of Standards and later in industry (Scott at Westinghouse Electric Corp. and French at the International Nickel Co.) worked on quenching round bars and measuring the depth of martensite layers. A major breakthrough came when Walter P. Jominy and A.L. Boegehold at the Research Laboratories Division of General Motors Corp. developed a test for measuring the "hardenability" of any particular steel or heat of steel.

This test consisted of heating a one-inch-round by three-inch-long bar of alloy steel to the heat-treating temperature, inserting it into a fixture, and directing a stream of water onto one end of the hot bar until the entire sample was cooled from the hardening temperature. Two flats were then ground on opposite sides of the bar. One side was placed on the anvil of a hardness-testing machine and a series of hardness readings were taken every 1/16 of an inch on the other flat side, starting at 1/16 of an inch from the quenched end. These hardness readings were then plotted as a function of the distance from the quenched end. The resulting curve is known as the *end-quench hardenability curve* or simply the *hardenability curve* (**FIG. 5.7**). It is the metallurgical equivalent of the "fingerprint" of a steel. It is the permanent identification of a given composition for a given set of heat-treating variables. Alloy steels are bought and sold today based on their "hardenability" as well as their chemical composition, and the "Jominy End Quench Test" is performed hundreds of times daily in steel mills and manufacturing plants all over the world to ensure their proper heat-treat performance.

At the same symposium in 1938 where Jominy reported on his test for hardenability, Marcus Grossman, Morris Asimow, and S.F. Urban published a paper, "Hardenability, Its Relation to Quenching and some Qualitative Data in Hardenability of Alloy Steel" (Ref 5). This research paper presented a mathematical analysis that allowed the true hardenability of a steel to be separated from the severity of the quench. Much later, Bain stated in his autobiographical book, *Pioneering in Steel Research*, "This monumental undertaking that grew out of the collaboration of the eminent metallurgist, Grossman, aided by the experimental research expert, Urban, with the distinguished mathematician, Asimow, stands as one of the few superlative scientific contributions to the science of

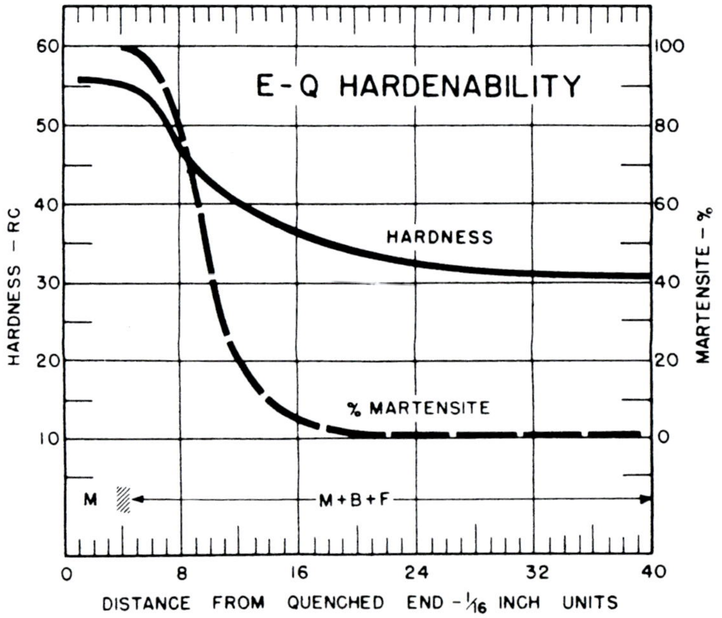

FIG. 5.7 Hardenability curve for SAE 4140 steel.

metals during the middle half-century (about 1915 to 1965)." Later, Grossman was able to calculate the contribution of each alloying element and elements in combination on the hardenability of alloy steel. He showed that the addition of small amounts of several elements was more effective than a larger amount of a single element. This multiplying effect allowed steels to be tailored to fit the exact requirements of the designed parts.

The Tempering of As-Quenched Martensite

The heat treating of alloy steels requires a reheating after quenching to a low temperature (usually 260–480 °C or 500–900 °F) called tempering. The as-quenched martensite is too brittle for any engineering application. Tempering is a complex process that has required considerable scientific study to reveal the underlying principles. The effect of alloying elements can alter the properties, especially toughness, available for any given set of tempering conditions. The properties of many alloy steels after tempering are reported in the book *U.S.S. Carilloy Steels, Alloy Steels for the Special Jobs of Industry,* published by U.S. Steel Corp. in 1948 (Ref 6).

Hardened alloy steel serves humankind in applications where high strength is needed. While ordinary steel, which industry calls "mild steel" or "carbon steel," can be used for bridges, buildings, ships, and most static structures, our machines require ever-increasing strength as more and more power is being transmitted in ever smaller packages. The breakthrough in understanding alloy steels has been very important to the development of the American technological age. All of this new science was in place in time for America to design a new series of alloy steels called "national emergency steels" to minimize the use of critical alloying elements for the massive use of alloy steels in World War II.

REFERENCES

1. J. Baur, U.S. Patent 49495, Chrome Steel Company of Brooklyn, 1865
2. E. Haynes, Materials for Automobiles, *ASME Trans.*, Vol 29 (No. 1148), 1907, p 209–16
3. W.R. Chapin, Properties of Some Steels in the Hardening Range, *Trans. Am. Soc. Steel Treating*, Vol 2, 1922, p 507–15
4. E.S. Davenport and E.C. Bain, Transformation of Austenite at Constant Subcritical Temperatures, *Trans. Am. Instit. Min. Met. Eng.*, Vol 90, 1930, p 117–54
5. M. Grossman, M. Asimow, and S.F. Urban, Hardenability, its Relation to Quenching and Some Qualitative Data in Hardenability of Alloy Steel, 1938
6. United States Steel Corporation, *U.S.S. Carilloy Steels: Alloy Steels for the Special Jobs of Industry*, Carnegie-Illinois Steel Corporation, 1948

Selected References

E.C. Bain, *Functions of the Alloying Elements in Steel*, American Society for Metals, 1939

E.C. Bain, *Pioneering in Steel Research: A Personal Record*, M.R. Hyslop, Ed., American Society for Metals, 1975

W.R. Chapin, Properties of Some Steels in the Hardening Range, *Trans. Am. Soc. Steel Treating*, Vol 2, 1922, p 507–15

E.S. Davenport and E.C. Bain, Transformation of Austenite at Constant Subcritical Temperatures, *Trans. Am. Instit. Min. Metall. Eng.*, Vol 90, 1930, p 117–54

M.A. Grossman, *The Principles of Heat Treatment*, American Society for Metals, 1935

H.M. Howe, *The Metallurgy of Steel*, Vol 1, The Scientific Publishing Company, 1892

W.A. Johnson and R.F. Mehl, Reaction Kinetics in Processes of Nucleation and Growth, *Trans. Am. Instit. Min. Metall. Eng.*, Vol 135, 1939, p 416–42

W.E. Jominy, *Hardenability of Alloy Steels*, American Society for Metals, 1939

R.R. Mehl, The Structure and Rate of Formation of Pearlite, *Trans. Am. Soc. Met.*, Vol 29, 1941, p 813–62

T.J. Misa, *A Nation of Steel: The Making of Modern America 1865–1925*, The Johns Hopkins University Press, 1995

P. Payson, Tempered Alloyed Martensites, *Trans. Am. Soc. Met.*, Vol 51, 1959, p 60

A.M. Portevin and M. Garvin, The Experimental Investigation of the Influence of the Rate of Cooling in the Hardening of Carbon Steels, *J. Iron Steel Instit.*, Vol 99, 1919, p 469

United States Steel Corporation, *U.S.S. Carilloy Steels: Alloy Steels for the Special Jobs of Industry*, Carnegie-Illinois Steel Corporation, Pittsburgh, 1948

Chapter Six

The Toolmakers

The progress of mankind has been marked by the ability to develop better tools. Simple tools were first devised for hunting, followed by tools for agriculture and construction, and tool making always included efforts to make better weapons for war. Finally, with the Industrial Revolution, humans began to invent mass-produced machine tools.

Early man built tools from materials at hand, including sticks, stones, and bones. Later, tools were shaped. They were sharpened to make suitable sticks for spears, and stones were shaped to make points and sharp edges. Eventually, man learned to reduce small amounts of iron from its ore through heat treating. This iron was a very primitive material. The early metal makers could not get their furnaces hot enough to melt the resulting mass that collected at the bottom, but this part-metal, part-furnace residue was able to be hammered, while hot, into useful shapes. In 1200 BC, the Iron Age began in Asia Minor; the dominant tool-making material was iron, supplying many of the tools and weapons in the Western world. The blacksmiths, the men who worked bars of iron into useful products, learned that heating iron in a bed of hot charcoal for a long time produced a layer of much harder iron on the surface of the tool they were making. If the hot metal was quickly plunged into water, they discovered, the surface layer could be made extremely hard. This process by these early blacksmiths was in fact allowing a layer of steel to form as the iron absorbed the carbon from the burning charcoal.

This was the beginning of the ultimate tool, made from heat-treated steel. Even though the techniques were very basic, a sword, knife, or spear made in this manner could be as useful or deadly as any similar object produced today.

The finest tools of all were the Damascus swords and Toledo blades of the Western world and the samurai swords of Japan. These were made by repeated operations of heating in charcoal and hammering out to long pieces that were folded back on themselves. With each operation, the two mating surfaces were welded together, eventually forming a structure made up of hundreds or even thousands of layers. Thus, the original thin layer of surface steel plus the new layers added with each reheating eventually produced a tool with carbon distributed throughout.

The Age of Making Steel

The earliest record of making steel by melting wrought iron and carbon together is credited to Aristotle, who in his wanderings came in contact with what was known as wootz steel, from India. Aristotle cited 350 BC as the date when wootz steel was first made, although some historians believe it may have been made earlier in China. Wootz steel was man-made from a simple, unsophisticated smelting furnace; it was worked into bars and cut into pieces. The small pieces were placed in clay crucibles to which wood chips were added. After sealing, the crucibles were heated in small furnaces to a high temperature by the use of an air blast from skin bellows. Melting occurred as the iron absorbed carbon from the wood. Once completely molten, the resulting steel was uniform in carbon. This was the first truly homogeneous steel.

The art of making steel existed in ancient and medieval times, but there was little technical knowledge or understanding. How the charcoal fire turned iron into steel was a complete technical mystery. Any explanations probably rested on the fire itself or, in the case of hardened steel, in the art of quenching.

Carburized iron, iron that is heated in contact with charcoal or other carbonaceous material to produce a surface layer of steel, was manufactured much more efficiently during the Middle Ages by processing a quantity of bars at one time in muffle furnaces, where combustion takes place in a separate chamber from the metal. The resulting product was known as blister steel due to the scale formed during the long heating cycle. Blister steel was hammered into tools, but the thin carburized layer did not lend itself to extensive hot forming or strong, heavy tools. To overcome this difficulty, the early steelmaker would forge carburized bars, cut them into short lengths, stack them, reheat, and forge again. The product of this process was called shear steel, and if the process was repeated, it was called double shear steel. In this manner, the original surface layers of carbon were distributed throughout the bar in thin bands to provide a more uniform material.

FIG. 6.1 *Teeming* by Robin Bell, in honor of clockmaker Benjamin Huntsman. A group of foundrymen cast molten steel from a crucible into a mold. Source: Malcolm Campbell.

Crucible Melting to Make Steel

During the middle of the 18th century, a British clockmaker named Benjamin Huntsman was looking for a uniform steel for clock springs when he rediscovered the ancient wootz method of melting small quantities of iron (blister steel in this instance, which had already been carburized) in clay pots (**FIG. 6.1**). For over 175 years (1740 to 1920), this process was used for making steels for tools (**FIG. 6.2**). The crucible process, as it was called, was suited for making small batches of high-quality, controlled chemistry steel—especially for adding the alloying elements that were found to improve tool performance.

The tool steels made in the first 100 years after crucible melting became available were simple iron-carbon alloys. During the 1860s, Robert Mushet was doing practical processing work on adding other metals to tool steel. His suggestion to Henry Bessemer, the inventor of a process for blowing air into molten iron, that he add a small amount of manganese to his steel after blowing air was critical to the success of the Bessemer process. Mushet's studies led him to the discovery (1868) that tungsten and manganese additions in sufficient amounts caused steel to be extremely hard upon cooling in air from a red heat.

FIG. 6.2 Crucible melting furnace, circa 1829. This is the oldest example of the Benjamin Huntsman process.

Conventional wisdom was that hardening could be done only by rapidly quenching in water. Mushet's discovery of adding alloys to steel to affect the way they could be hardened was a historic milestone in the story of tool making (**FIG. 6.3**). His technical discovery was so little understood, however, that it did not lead to further advances at the time. The company he formed to make his "special steel" failed. Luckily, the steel survived, made by an established Sheffield crucible steelmaker, and it was still the principal tool steel in use as late as 1900. The steel contained 2% carbon, 2.5% manganese, and 7% tungsten. It is considered the ancestor of all modern tool steel. It could withstand greater service conditions and required less sharpening due to its greater resistance to wear.

The locale of the development of superior steels for tools shifted momentarily to the Jacob Holtzer Steel Company in Loire, France. Henri-Aimé Brustlein

FIG. 6.3 Robert Forester Mushet, inventor of the first alloy tool steel.

was studying the use of chromium as an alloy in steel. Brustlein's interest in tool steels occurred around 1890, when he became involved in using chromium as a replacement for manganese in the tungsten tool steels of Mushet. He brought samples of his steels to the United States in 1892 and discussed their advantages with the Midvale Steel Company in Philadelphia. At this critical time, Frederick W. Taylor was conducting studies on production efficiency at Midvale Steel (Ref 1). He was conducting pioneering work on time and motion measurements, and he became world famous as the founder of the "efficiency expert" concept (**FIG. 6.4**). During this period, Taylor became interested in time studies on machining metals. Thus, he became familiar with the new Brustlein-type tool steels. Later he went to work for the Bethlehem Steel Corporation, where he was again concerned with machining efficiency. Taylor, with the help of an experienced Bethlehem metallurgist, Maunsel White, undertook a study of the heat-treating variables affecting the performance of alloy tool steels. Their results revolutionized the modern tool steel industry.

In the course of their study, Taylor and White found that the higher the heating temperature before the steel was cooled, the greater the usefulness and life of the tool. Metalworkers had known all these years that if steel was heated at too high a temperature it would be ruined, or "burnt," as it was described. However, Taylor and White followed the clues provided by their experimental work and found that the higher the temperatures used, the more useful the tool. Alloy steel tools containing tungsten were so greatly improved if heated to

FIG. 6.4 Fredrick Winslow Taylor (pictured) and Maunsel White discovered the heat treatment that produced secondary hardening in tungsten tool steels.

these extremely high temperatures before cooling that they could be operated under conditions so severe the cutting point would glow to a dull red. Later, with some refinement in alloy content, these would be called high-speed steels, and they would be said to possess "red hardness." The discovery of high-speed steels, heat treated by the Taylor and White method, revolutionized the machine tool industry just when America was gearing up for the age of mechanization (**FIG. 6.5**).

The crucible melting process was still being used to make high-speed steel, but by 1906, the Sanderson Brothers Steel Company of Syracuse, New York, installed the first electric arc furnace in the United States for melting tool steels. Other tool steel companies were formed about this time, and electric arc melting was adopted as the preferred method. As the tool steel industry expanded in the early decades of the 20th century, many alloy types were invented or imported from England and Europe. In these early years, each steel company used its own brand names, numbers, or other special identifications for many different grades of steel, designed for a wide variety of applications such as hand tools, cold-forming dies, hot-forming dies, lathe tools, and milling cutters. Many of these steels were higher-carbon (for higher strength or hardness)

FIG. 6.5 A typical lathe before the use of high-speed steel. The lightweight lathe had to be replaced to withstand the heavy cuts now possible with high-speed steel tools.

versions of early alloy steels from France, Germany, and England. Secrecy, rather than patent coverage and facts, was resorted to in making and marketing tool steels. This state of affairs was confusing to the users of these steels. However, it provided a method of bonding the user to the producer in the belief that he would obtain consistent performance.

Taylor and White did their initial work in 1898, and their new knowledge was demonstrated at the Paris Exposition of 1900. The only presentation on this great achievement was an address by Taylor at the American Society of Mechanical Engineers in 1906. Taylor and White, and their employer, realized from the beginning that they had not invented a new steel, but rather had developed a breakthrough in alloy steel heat treating. Patents were applied for and obtained on this unique, extremely high-temperature 1205 to 1315 °C (2200 to 2400 °F) treatment for hardening. The commercial goal of the Bethlehem Steel

Corporation was to license manufacturers in using this treatment. The commercial strategy soon disintegrated because many shops simply ignored the patents; when Bethlehem went to court to enforce its patent, it lost the long, drawn-out court battle.

New Alloy for Tool Steels

In the meantime, improved alloys that would give maximum response to the Taylor and White heat treatment were being tested in many of the advanced industrial countries. A young metallurgist who was active throughout these times was J.A. Mathews of the Crucible Steel Company. Mathews reported the results of a survey that he took in 1901 on the common tool steels in use. "Modern high-speed steels seem to have sprung fairly fully developed from a variety of sources at almost the same time," he wrote. He further stated that he had been unable to show just when the change from the old type of high-speed steel took place. The change from the old type to the new that Mathews was referring to was a rather drastic change in chemistry from the old Mushet steels, containing 1.5 to 2.0% carbon, 2.5 to 4% manganese and 7 to 9% tungsten, to new steels with 0.6 to 0.8% carbon, 4 to 6% chromium (manganese was no longer used), and 10 to 20% tungsten.

Quoting from Mathews, "It is probable that we shall never know who took the radical steps and made the first low-carbon, high-tungsten, high-speed steel, but immediately following the announcement of the Taylor-White process there was great activity on the part of all tool steelmakers in every country to produce a product which would yield maximum results when treated by this process. The courts decided that Messrs. Taylor and White did not make a patentable invention or discovery. The world, however, seemed to differ with this decision and every scientific honor was awarded them for their epoch-making announcement which revolutionized machine shop operation and machine-tool building."

Mathews was more than just a casual observer on the high-speed steel scene. His patent, granted on January 3, 1905, for the addition of vanadium to high-speed steel was discussed in his own words as follows: "This writer began experimenting with the use of vanadium in 1903, and it is well to bear in mind that at that time vanadium was almost a chemical curiosity. It was worth about $15 a pound, and this was some time prior to the formation of the American Vanadium Company that manufactured and sold vanadium in large quantities. So far as the writer is aware, the entire stock of ferrovanadium in the country when these experiments were begun consisted of not over 100 pounds in the hands of two different dealers in New York. We purchased half of the entire stock of each dealer."

The experimental work described by Mathews was carried out at the old Sanderson Works of Crucible at Syracuse, New York. Mathews goes on to say that J.M. Gledhill of England and Taylor himself had discussed the use of vanadium in 1904. With the addition of about 1% vanadium to the 18% tungsten, 4% chromium, and 0.60 to 0.80% carbon steel, the first truly universal high-speed steel was born (**FIG. 6.6**).

The fact that no one had knowledge as to why this type of steel exhibited such unusual behavior in service did not prevent the researchers from maximizing the service performance to such an extent that even today this 18-4-1 high-speed steel (known commercially in the United States as T-1) is still in limited commercial service. By 1905, the new high-speed steel was in commercial production and would remain the major tool steel for metal machining for the next 35 to 40 years.

A great deal of practical development work was done in the first 10 to 15 years after the initial discoveries. Cutting tests were performed by tool steel producers and users to optimize the many variables of alloy content, hardening temperatures, methods of cooling from hardening, and post-hardening heat treatments (tempering). Not all of this work was performed with equal competence or similar results. Thus, the times spawned many ideas regarding what constituted the ideal tool material. This situation was further compounded by a lack of understanding the metallurgy involved in such high-alloy steels.

FIG. 6.6 Tools made with 18-4-1 high-speed steel.

Research on High-Speed Steel

H.C.H. Carpenter of the National Physical Laboratory (Great Britain) reported on one of the earliest research investigations of the metallurgy of high-speed steels in 1905 and 1906 (Ref 2, 3). He studied the microstructural changes that occurred in many different English and American commercial steels as well as in controlled laboratory alloys, in which he varied the chromium, tungsten, and molybdenum contents.

Carpenter, however, correctly concluded that Mo and W were the elements that provided the resistance to softening. He cites Henry Louis Le Chatelier, the famous French scientist, as previously arriving at this same conclusion. He misunderstood the important role of Cr in providing the air hardening. He also believed that the high temperature phase (austenite) was the source of red hardness. He was led into this error due to the large amount of austenite he saw in metallographic samples after cooling from the hardening temperatures. However, Carpenter did realize that resistance to tempering was a separate subject worthy of study. The next year (1906) he performed tempering studies but did not shed any new light on the red hardness phenomenon with the limited tools of this time.

Another metallurgical research study was undertaken by a countryman of Carpenter's two years later (1908). C.A. Edwards, a Fellow at Victoria University at Manchester, studied the effect of Cr and W and considered hardness itself as a contributor to red hardness. Taylor had always thought of red hardness as a separate special property that could be studied only by a cutting test. While Edwards' first work did little to improve the understanding of high-speed steel metallurgy, he later collaborated with H. Kikkawa (1915) to produce a landmark paper on the subject.

They performed hardness tests in a systematic study of the effect of Cr and W on tempering. This method showed the hardness peaking after tempering at 540 to 650 °C (1000 to 1200 °F) (**FIG. 6.7**). In fact, using a steel similar to Taylor's recommendation, they showed a secondary hardening peak at just the temperature that Taylor had predicted in his work 15 years earlier.

Edwards and Kikkawa provided the first comprehensive understanding of the major metallurgical phenomena in high-speed steel. They concluded that Cr imparts the self-hardening and that the extremely high temperature for hardening was needed to dissolve the tungsten carbides. The maximum resistance to tempering (or maximum secondary hardening peak) could only be obtained by getting the tungsten into solution. They also concluded that careful tempering studies with hardness measurements could provide valuable information on the relative merits of cutting tools (expensive and time-consuming cutting tests had been the only reliable criterion up to this time). Interestingly, the tempering of

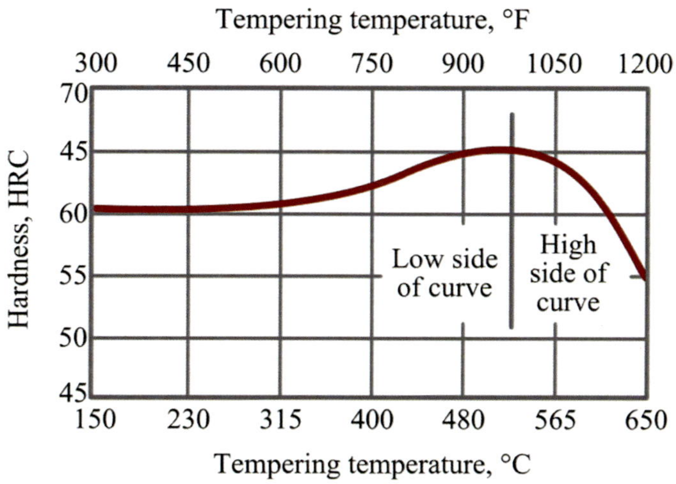

FIG. 6.7 A high-speed steel tempering curve showing the hardness peak at Rockwell 65.

high-speed tools at 565 to 620 °C (1050 to 1150 °F) was not recognized as being of prime importance until after World War I.

Advanced Research on High-Speed Steel

Shortly after Edgar Bain's early work using x-rays to determine the crystal structures of austenite (fcc) and martensite (bcc), he and Zay Jeffries published their famous paper in *Iron Age* in 1923 on the "Cause of Red Hardness of High Speed Steel," (Ref 4). This paper is a classic in the field of metals technology, not because it changed industrial practice but because it combined the latest research tool (x-ray diffraction) with the latest theory of hardening (slip interference by precipitated particles). Bain performed the x-ray work and Jeffries supplied the slip interference concept.

This paper showed, as concluded earlier by Edwards and Kikkawa, that the high hardening temperatures are needed to dissolve the particles of tungsten-containing carbide in the austenite. Bain and Jeffries then concluded that the softening of hardened steel during tempering, which occurs in ordinary steel at low temperatures of 150 to 480 °C (300 to 900 °F), is caused by grain growth and carbide particle growth beyond the critical size. They reasoned that the greater stability of the tungsten carbide forces its formation at the higher temperature. It is only at these temperatures that the larger tungsten atoms can move within the metal space lattice to form the alloy carbides. It would be many years before most of the concepts proposed by Bain and Jeffries could be properly evaluated. Later studies would show they were correct in the thrust of their theories, but the details of alloy carbide formation would be more complex.

The following year, 1924, Edgar Bain moved from General Electric to Atlas Steels in Dunkirk, New York. He worked with one of America's most interesting and prolific metallurgists, Marcus A. Grossman, while there. The publications of Bain and Grossman in 1924 included high-carbon chromium steels, chromium in high-speed steel, and their major work, "On the Nature of High Speed Steel," which they published in Great Britain (*Journal of the Iron and Steel Institute*) rather than in the United States (Ref 5). This paper was a compilation of the arts involved in the manufacture and metallurgy of high-speed steel. Grossman and Bain expanded this effort in their collaboration in 1931 with the publication of a textbook entitled *High Speed Steel* (Ref 6).

Molybdenum as a Substitute for Tungsten

A shortage of tungsten during World War I forced many users of high-speed steels to fall back to using carbon steels, a regression of 50 years in technology. After the war, a major research effort in high-speed steels was to substitute molybdenum (the sister metal for tungsten) in the T-1 alloy of Mathews. Work was done at Watertown Arsenal in Massachusetts during the late 1920s and early 1930s, where the approach was to substitute about 9.5% Mo for 18% W in T-1. This was roughly a one-for-one atomic substitution because Mo has only one-half the atomic weight of W, so only about one-half weight percent was needed. Major accomplishments included the use of a borax coating during heat treating to protect the surface and the use of molten salt baths for heating steels for hardening. Drawbacks on all prior high-molybdenum tool steels had been the excessive loss of carbon at the surface during heating.

It was only natural that an army arsenal would be vitally interested in developing Mo high-speed steels. The United States was dependent upon other countries, especially in Asia, for the bulk of its tungsten. Shortages during the war showed the strategic need for finding a substitute. The United States did not have much tungsten, but there were large deposits of molybdenum in the mountains of Colorado.

A small mining and refining venture was started by several of the principals of the American Metals Company in 1918 to produce molybdenum for alloy additions in armaments such as tanks and guns for World War I. This activity took the name of a local railroad station near Climax, Colorado. The mine was shut down completely after the war due to limited demand. Within a few years, however, consumption increased at a rapid pace, aided by the increasing use of alloy steels in general, but especially in the rapidly growing auto industry. With increased supplies from Climax Molybdenum Company, the price of the metal was finally reduced until it became competitive with tungsten for high-speed steel.

Price, however, was not the only deterrent to replacing tungsten with molybdenum. Technical acceptance in the industrial marketplace among the tens of thousands of toolmakers, tool-room foremen, and machinists was the final demand that had to be satisfied. The Watertown Arsenal research found many applications for the molybdenum (tungsten-free) high-speed steel within the various arsenals around the country, but molybdenum never gained acceptance in industry as a competitor of T-1.

Cleveland Twist Drill Company is First Adopter of Molybdenum High-Speed Steel

The first significant breakthrough in the commercial development of Mo high-speed steels was work performed by Joseph V. Emmons at the Cleveland Twist Drill Company. His work was undertaken in the late 1920s and continued through the dark economic times of the 1930s. In 1933, Emmons received his patent and published a technical paper.

Both his paper and patent were models of clarity and precision. Emmons reported that all-molybdenum (i.e., tungsten-free) steels were, in fact, inferior to T-1. He also pointed out that substituting small amounts of molybdenum for some of the tungsten was not worthwhile. But he made a major discovery that a ratio of about four parts Mo to one part W, for a total of 10% of the steel, provided a critical composition that could compete against the 18% tungsten of T-1. His paper entitled “Some Molybdenum High Speed Steels” was presented in Buffalo before the national conference of ASM in October 1932 (Ref 7). It won the coveted Henry Marion Howe Medal for that year.

Later, interference was declared by the Patent Office between Emmons’ application and one by Frank Garratt of Universal Steel Company. The problem was resolved when Garratt gave priority to Emmons, and Cleveland Twist Drill licensed Universal to make the steel. Patent No. 1,937,334 was issued to Emmons on November 28, 1933.

By the end of the 1930s, patents by Emmons, James P. Gill at Vanadium-Alloys Steel Company, and others provided a sound basis for commercial development of the new steels when the tungsten shortage occurred again during World War II. In fact, the Cleveland Twist Drill Company had already provided a significant body of experience in this commercial work through its use of these steels under the leadership of Emmons and in cooperation with the other tool-steel producers, especially Crucible and Universal. Industrial practices, however, are difficult to change, and at the time of the United States’ entry into World War II, over 80% of the high-speed steels were still the tungsten type.

The War Production Board provided encouragement for the massive transfer to molybdenum by denying the tool steel industry any tungsten to maintain production. Thus, molybdenum high-speed steels became the dominant type during World War II and after.

MIT Research on High-Speed Tool Steels

It was now 40 years since the discovery of the high-speed steel heat treatment by Taylor and White and 15 years since Bain and Jeffries published their theory of secondary hardening. While practical development of the Mo high-speed steels was accomplished during this time, little was done to develop a more fundamental understanding of the materials. A young professor of metallurgy at the Massachusetts Institute of Technology (MIT), Morris Cohen, set about to develop an understanding of the nature of the tempering of high-speed steel (**FIG. 6.8**). Starting in 1939, studies were published by Cohen and a series of graduate students who did their doctoral theses on high-speed steel.

The first paper by Cohen and P.K. Koh (1939) set the tone for most of the papers that followed (Ref 8). They reviewed the literature on secondary

FIG. 6.8 Morris Cohen and graduate students published a series of papers on tempering of high-speed steel.

hardening, which showed that most previous researchers believed the high-speed phenomenon was caused by the transformation of residual austenite to martensite during tempering. Considerable so-called residual austenite was always seen in the microstructure of as-cooled high-speed steel, but it was replaced by martensite after tempering. Cohen and Koh studied changes in properties after heat treating at various temperatures and times. The variable of time at tempering temperature had never been examined adequately. In addition to x-ray diffraction on solid samples, they studied changes in electrical and magnetic properties, length, volume, and hardness. They concluded that there were four stages to the reactions in high-speed steel during tempering:

1. Formation of iron carbide (Fe_3C)
2. Precipitation of carbide in the retained austenite
3. Transformation of retained austenite to martensite
4. Precipitation of alloy carbides in martensite

The Cohen and Koh paper showed for the first time in American technical literature that the retained austenite did not transform to martensite at the tempering temperature, but on cooling from this temperature. Time spent at the tempering temperature conditioned the residual austenite for this subsequent transformation. The basic conclusion was that stages two and three contributed to secondary hardening. This conclusion continued the conventional wisdom that somehow the transformation of retained austenite was the cause of secondary hardening, or red hardness, and that the eventual formation of alloy carbides upon long exposure to temperatures of 595 °C (1100 °F) or higher was not significant. This initial paper by Cohen and Koh was of practical importance to the tool steel heat treaters as well as being a document concerned with the theory of secondary hardening (Ref 8).

Over the next decade, Cohen and his students continued their research studies on steel. Some work was directly applicable to high-speed steel, while other work was more basic to steels in general. The work on tempering high-speed steel showed that little was understood about the tempering of all hardenable steels. Such a study was undertaken by Antia, Fletcher, and Cohen and reported in 1944. One of the last papers in the high-speed steel series (Cohen and Blickwede) was on the effects of vanadium and of carbon on a 6% tungsten–5% molybdenum high-speed steel. This new steel was becoming very popular in industry under the designation M2 (**FIG. 6.9**). The study was performed on the full range of carbon and vanadium contents that would ever be useful in high-speed steel. This work was important to later research on M2 to produce ultra-high hardness steels.

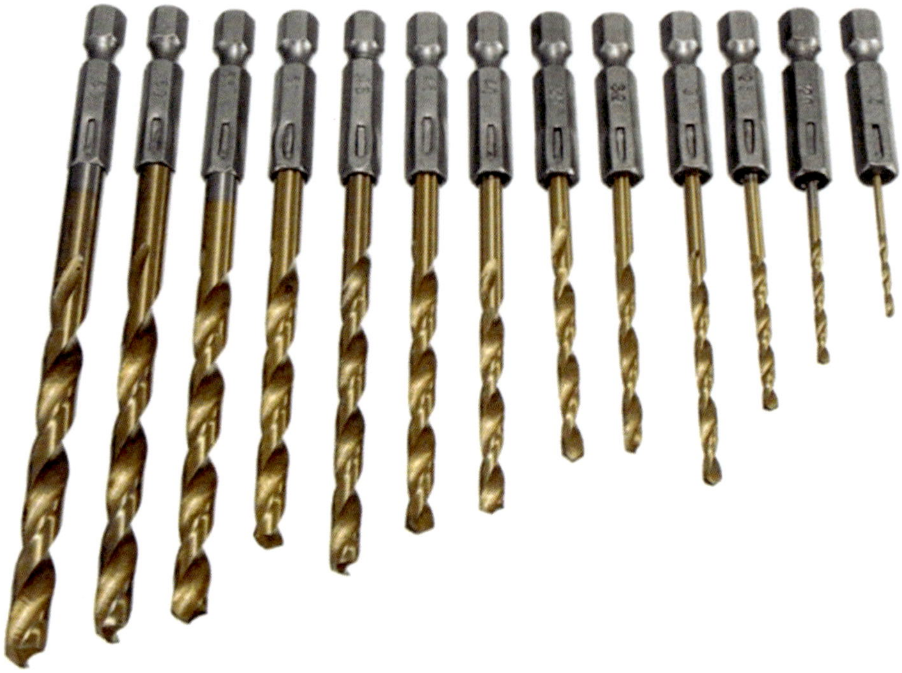

FIG. 6.9 Modern high-speed drills made of M2 and coated with titanium nitride for added wear resistance.

Research at the Union Carbide and Carbon Laboratories

Walter Crafts and John Lamont of the Union Carbide and Carbon Research Laboratories published a paper in 1948 that was of importance to high-speed steel. Union Carbide and Carbon Corporation was a major supplier of ferroalloys, especially ferrochrome, to the steel industry. They built one of the earliest metallurgical laboratories in this country (first on Long Island and later at Niagara Falls, New York) and made many contributions in stainless and alloy steels. Crafts and Lamont were studying the effects of alloys in ordinary engineering-type steels. They had increased the amount of alloy until they were approaching the alloy content of tool steels. They found secondary hardening peaks after tempering, even though their steels did not contain residual austenite after quenching. This work finally killed the retained-austenite theory, so stubbornly held since the days of Carpenter over 40 years earlier.

Even more important was their work on the newly developed electron microscope. By the 1940s, the electron microscope was being developed. This microscope used a stream of electrons, rather than light rays, to see objects that were many times smaller than could previously be detected. Although

Crafts's and Lamont's work was done at magnifications only about twice that available by light microscopes, they found tiny carbide particles forming at tempering temperatures corresponding to the high-speed steel hardening peak. These small carbides were found by x-ray diffraction to be the alloy carbides of Mo_2C, W_2C, or VC in various Mo, W, or V steels. Crafts and Lamont were pioneers in using the electron microscope to study carbide precipitation in alloy steels. This initial work supported the 25-year-old theory of Jeffries and Bain that alloy carbides were the direct cause of secondary hardening in high-speed steels.

Advanced Work in Sweden

Kehsin Kuo, a research chemist who was born in Beijing and educated at the University of Chekiang, China, traveled to Sweden in 1949 to study metallography under Professor Emeritus Axel Hultgren at the Royal Institute of Technology, Stockholm. Later he moved to the University of Uppsala to study carbides under Professor Gunnar Hägg. These two Swedish researchers had international reputations in their respective fields, and the result was a series of research papers published in the British *Journal of the Iron and Steel Institute.*

Kuo's first major work was a study using the latest techniques of x-ray diffraction on the carbides that form in steels alloyed only with chromium, molybdenum, or tungsten. As part of this work, he examined the carbides that form in commercial high-speed steel. Included in the commercial steels were the old standby T-1, as well as M2 and several European steels, along with some of these same steels with cobalt additions. Kuo reported that the first alloy carbide to form when tempering high-speed steels in the temperature range where secondary hardening occurred was W_2C for high-tungsten steels and Mo_2C for molybdenum steels. At still higher temperatures, where the hardness decreased rapidly, the W_6C and Mo_6C carbides were found. Because tungsten and molybdenum are sister metals that formed the same type of carbides, he found they could be combined, in which case the carbide becomes M_2C or M_6C, where M stands for metal atom tungsten, molybdenum, or a combination of them. Kuo concluded that the formation of this M_2C carbide was the cause of secondary hardening (and red hardness) and that the transformation of retained austenite was not needed to explain the increased hardness. Using the latest techniques of the electron microscope, Kuo later showed that the M_2C carbide first precipitates as extremely fine, threadlike particles that are visible at magnifications of 100,000 times their actual size.

Developments during the 1950s and 1960s were largely in the replacement of T-1 and its relatives with M1, M2, M7, M10, and other molybdenum grades. Research work shortly after World War II by Crafts and Lamont at Union Carbide and Carbon Research Labs led to a concept of a balanced composition in

FIG. 6.10 A modern metalworking lathe with computer numerical control using high-speed steel and high-speed steel with cobalt for the tools.

the tempering of alloy steels. This concept was based on adjusting the carbon and alloy content to supply just the right balance for fulfilling the needs of the various carbides that formed during the critical stage in tempering. Additional studies produced a series of steels that optimized the composition of previous high-speed steels by increasing the carbon content to fulfill the balanced designed concept and adding 5 to 8% cobalt, obtaining the highest hardness by the C scale of the Rockwell Hardness Test (RC 68) ever consistently obtained in high-speed steels. Cobalt had been used in high-speed steels since the earliest days, but sound laboratory evidence of its value was not available until this recent work. These new cobalt high-speed steels could machine harder materials, notably the jet engine superalloys and space-age metals being developed at this time (**FIG. 6.10**).

More than 150 years since Robert Mushet invented the first alloy tool steel, and over 100 years after the discovery of the special heat treatment by Taylor and White, the modern technological society came into being, in no small part due to the availability of modern cutting tools with high-speed steel.

REFERENCES

1. F.W. Taylor, The Heat Treatment of Tungsten Steel, *Trans. Am. Soc. Mech. Eng.*, Vol 31, 1907
2. H.C.H. Carpenter, Tungsten in High-Speed Steel, *J. Iron and Steel Instit.*, Iron and Steel Institute, 1905
3. H.C.H. Carpenter, The Heat Treatment of High-Speed Steel, *J. Iron and Steel Instit.*, Iron and Steel Institute, 1906
4. E.C. Bain and Z. Jeffries, The Cause of Red Hardness of High-Speed Steel, *Iron Age*, Vol 112, 1923, p 805–810
5. M.A. Grossman, On the Nature of High-Speed Steel, *J. Iron and Steel Instit.*, Vol 110, 1924
6. M.A. Grossman and E.C. Bain, *High Speed Steel*, John Wiley & Sons, Inc., New York, 1931
7. J.V. Emmons, Some Molybdenum High-Speed Steels, *Trans. Am. Soc. Met.*, Vol 21, 1933
8. M. Cohen and P.K. Koh, “Secondary Hardening in High-Speed Steel,” *Trans. Am. Soc. Met.*, Vol 27, 1939

Selected References

C.S. Smith, Ed., *Sorby Centennial Symposium on the History of Metallurgy*, Gordon and Breach Science Publishers, New York, 1963

Chapter Seven

Stainless Steel

One of the greatest advances in modern metallurgy was the discovery of a steel that did not rust. Until this point, all steel would eventually transform to iron oxide (i.e., rust). The discovery of this new steel was made when research started on high-chromium alloy steels. The addition of chromium or nickel in regular heat treated alloy steels was done in France during the 1880s and 1890s, and chromium combined with nickel was being produced in Germany in the 1890s. The alloy additions were in the 2 to 5% range, and carbon was needed to heat treat these steels to higher strengths.

Before 1895, there was no process for producing low-carbon chromium to be added to iron, an important factor in chromium steels. At this time, German chemist Hans Goldschmidt invented the thermite process, a process that combines chromium ore with aluminum powder to generate enough heat to separate the chromium metal without carbon contamination. This paved the way for the study of alloys with much higher chromium content and low carbon.

The first metals researcher to make use of this new low-carbon chromium was a Frenchman named Leon Guillet (**FIG. 7.1**). From 1904 to 1906, he published his research, which covered the full range of chromium additions that would later be alloys within the commercial range of stainless steel. Guillet's work showed three basic alloy types. One was low carbon with high chromium that could not be hardened; this was ferritic. Another had higher carbon that could be hardened just like any other alloy steel; this was martensitic. Finally, he studied alloys with both chromium and nickel. With sufficient nickel, these alloys were austenitic. This was the first time an all-austenitic microstructure was seen at ambient temperature.

FIG. 7.1 Leon B. Guillet studied high-chromium steels for the first time.

These three categories define the range of stainless steels even today. His studies covered the metallographic and mechanical properties that could be measured at his time. However, Guillet missed the fact that some of the alloys were resistant to corrosion in acids and at normal atmosphere. Despite this, Guillet was the first to publish information about alloys that later became the stainless steels, and because of this, Carl Zapffe, in his classic *Stainless Steels* book (ASM International, 1949), awarded him first place for the discovery of these alloys (Ref 1).

Two other researchers who studied iron-chromium and iron-chromium-nickel alloys a few years later included the well-known French metallurgist Albert Portevin and an Englishman, W. Giesen. Portevin studied the straight iron-chromium alloys with both low and high carbon. Giesen did most of his work on the iron-chromium-nickel austenitic steels, which had to be low carbon. Both published their results in the Iron and Steel Institute's *Carnegie Scholarship Memoirs*, Vol. 1, 1909 (Ref 2).

The Discovery of Stainless Steel

The first researcher to find the most important property of resistance to corrosion was a German named Philip Monnartz (**FIG. 7.2**). He started his work in 1908 and published his findings in the journal *Metallurgie* in 1911 (Ref 3). Quoting from Zapffe, "His research disclosed the stainlessness of stainless steels—and provided as brilliant a piece of work as stands in the literature on stainless steels" (Ref 1). First, Monnartz determined that when the chromium content exceeded 12%, the steel was resistant to attack by nitric acid and did not corrode in the normal atmosphere. Next, he found that the metal surface needed to be prepared to form a protective film. This extremely thin film was the formation of an oxide with a combination of iron and chromium when the chromium exceeded 12%. The tightly adherent film prevented further oxidation or corrosion. He also found that low carbon was necessary to prevent the formation of chromium carbides and that the carbon could be controlled by adding stronger carbide formers, such as titanium, columbium, vanadium, molybdenum, or tungsten, to form carbides of these alloys rather than that of chromium. Finally, Monnartz learned that the addition of 2 to 3% molybdenum greatly increased the corrosion resistance of austenitic stainless. His research provided the basic information for all future developments in practical alloys for the many applications that would be found for stainless steel.

For his discovery, Monnartz must be considered the one true inventor of stainless steel. While the other researchers, especially Guillet, were the early

FIG. 7.2 Philip Monnartz was the first researcher to discover that alloy steels with more than 12% chromium were stainless.

workers on the iron-chromium and iron-chromium-nickel steels, they completely missed the only property that makes these alloys useful today. The mystery is how they did not notice that some of their alloys were so difficult to etch for metallographic examination. As stated by Zapffe, "It's not surprising that it took so long to discover stainless steel. What's surprising is that it wasn't discovered sooner."

Patents

The years from 1904 to 1911 were a major time of discovery, and the following decade saw the development of alloys for specific applications. The first patent for a stainless steel alloy was German No. 246,015, issued in 1910 to W. Borchers and P. Monnartz (Ref 4), but the first application for stainless steel was for lead-in wires for electric lamp bulbs. The research was done by Christian Dantsizen at the General Electric Research Laboratory in Schenectady, New York. Using very low-carbon chromium made by the Goldschmidt thermite process, he developed an alloy containing 14 to 15% chromium and 0.07 to 0.15% carbon. His goal was resistance to oxidation at high temperature, and the chromium provided that in a ferritic microstructure. GE used this alloy a few years later for steam turbine blades.

Another American, Elwood Haynes, worked on stainless steel that contained higher carbon content to produce a martensitic microstructure on quenching from a high temperature. Haynes was basically an automobile manufacturer with an interest in developing alloys for spark plugs (**FIG. 7.3**). He had already invented Stellite, a cobalt-chromium-tungsten-carbon alloy that was used to make very hard wear-resistant surfaces, such as valve seats. He applied for a patent in 1911 for a stainless steel alloy that contained 13% chromium with 0.30% carbon. His patent was held up for many years due to interference with a British patent.

The British patent was held by Harry Brearley, who had been searching for a wear- and corrosion-resistant alloy for gun barrels (**FIG. 7.4**). This led him to a 13% chromium, 0.25% carbon hardenable stainless steel. He promoted this alloy, the forerunner of AISI 420, for cutlery. The alloy was a natural fit for hardened knives, and his employers were in Sheffield, England, the capital of tool and knife making. Brearley was credited with starting the first production of hardenable stainless. The British military found that Brearley's alloy was the answer for aircraft engine exhaust valves and requisitioned the available supply for the war years 1914 to 1919. Thomas Firth and Sons made 50 tons for knives before the war. That production lot is considered the beginning of the stainless steel industry.

FIG. 7.3 Elwood Haynes, the American car manufacturer who invented the 12% chromium, 0.30% carbon hardenable stainless steel

FIG. 7.4 Harry Brearley invented the 12% chromium, 0.25% carbon steel used for cutlery by the Sheffield knife manufacturers.

One of the most important patents issued for stainless steel went to Eduard Maurer and Benno Strauss, who worked at the Friedrich Krupp Works in Essen, Germany. Maurer studied under Osmond at Paris. His thesis was on hardenable stainless steel and he was hired as the first metallurgist at the Krupp Steel Works. Their German patent No. 304159, issued in 1912, included, for the first time, an austenitic stainless steel with an alloy containing 20% chromium and 7% nickel (Ref 5). With slight modification, this alloy became the 18-8—the most important stainless steel alloy ever produced (**FIG. 7.5**). The years from 1910 to 1920 saw little advance in producing stainless steel. All available metal was diverted during the war to produce exhaust valves for aircraft engines in both Britain and the United States.

Some cutlery and tableware were produced in Sheffield, England, before the war. The first heat of stainless steel in the United States was melted at the Pittsburgh plant of the Firth-Sterling subsidiary of Thomas Firth and Sons in Sheffield. Another early producer was the Carpenter Steel Company, which produced material for the Liberty Aircraft engines in 1917. Most of the production was in the heat-treatable martensitic grade steel, 13% chromium and 0.30% carbon. The general martensitic grades were covered in the United States by

FIG. 7.5 Eduard Maurer, the first metallurgist to work at Krupp Steel Works in Germany.

patent applications made in March 1915 by both Elwood Haynes and Harry Brearley. Rather than fight the conflict in court, Brearley and Thomas Firth and Sons formed a syndicate named The Stainless Steel Company to hold all patents. Haynes agreed to a 30% share in the company, and Brearley and Firth held 40%. Several other U.S. companies held the remaining 30%. This solved the problem for ferritic and martensitic grades of stainless. Krupp in Germany held the patent rights for the austenitic grades.

Applications

Through most of the 1920s, only ferritic and martensitic stainless steels were made in the United States. The lack of the austenitic grades was corrected when patents were exchanged between England and Germany in 1923, and a royalty was paid to import or to produce austenitic grades in the United States in 1928.

The earliest uses after the war were again cutlery and tableware. By 1923, this expanded to surgical and dental instruments and then to containers for nitric acid. New uses found throughout the decade included milk-handling equipment, surgical implants, cookware, golf clubs, trim for autos, and the spectacular curtain wall for the upper seven stories of the new Chrysler Building in New York, including the huge gargoyles (**FIG. 7.6**). The Chrysler Building used 48 tons of the 53,000 tons of austenitic grade stainless produced in 1929. Uses for the austenitic grade continued to grow. Allegheny Steel, Ludlum Steel, Carpenter Steel, Crucible Steel, Lukens Steel, Latrobe Steel, and others were the major producers in forming the industry.

Meanwhile in England, W. H. Hatfield had modified the German austenitic alloy containing 20% chromium plus 7% nickel to the now familiar 18% chromium plus 8% nickel (18-8) that became AISI 302 in the United States. He also added titanium to combine with the carbon to form AISI 321. This alloying greatly improved weldability. The 18-8, the most important stainless steel alloy, had the ideal combination of formability with corrosion and acid resistance, and it could be polished to a beautiful surface finish.

The 1930s and 1940s

The Great Depression of the 1930s impacted stainless steel, as it did all metal making. Production decreased from 58,000 tons in 1929 to 23,000 tons in 1932. There were a few growing uses during the decade, mainly in transportation—automobiles, trucks, and trains. Automobiles used more trim and trucks were lined to carry milk, other foods, and acids. The most interesting use was trains. Budd Manufacturing Company designed the Burlington Zephyr, a passenger train made of stainless steel. The train consisted of the locomotive, baggage car,

FIG. 7.6 The Chrysler Building's upper seven stories are clad with the German alloy, 20% chromium and 7% nickel.

and two passenger cars. Aircraft-type construction was used. From 1935 to the start of World War II, 47 stainless steel trains were manufactured. The original Zephyr is on display at the Museum of Science and Industry in Chicago (**FIG. 7.7**). Another interesting use was for automobiles built by the Ford Motor Company using Allegheny stainless steel (**FIG. 7.8**). One of the main uses for stainless steel during the war was for aircraft exhaust systems and engine valves. Other uses included ship galleys and hardware wherever corrosion was a problem.

Development of New Alloys

Following World War II, several stainless steel producers looked for alloys that responded to precipitation hardening. The earliest of these alloys was Stainless W, developed by the Carnegie-Illinois Steel Corp. In 1948, a series of alloys were developed by the American Rolling Mill Company (Armco). They kept the

FIG. 7.7 Budd Manufacturing Company's first stainless steel train, the Burlington Zephyr.

FIG. 7.8 Three stainless autos built by the Ford Motor Company, the 1936 Ford, 1960 Thunderbird, and 1967 Lincoln. Source: Photo by Dean Shaw.

chromium content at the 17% level but reduced the nickel to 4% and added 3.5% copper (17-4PH) or 7% nickel with 1% aluminum (17-7PH), and made a third alloy with 7% nickel, 2.5% molybdenum, and 1% aluminum. These alloys had design strengths of 200,000 to 220,000 psi in the heat treated condition. Other stainless producers developed similar alloys. Another alloy system under study was the duplex stainless steels. The composition was designed to have a microstructure that was about 50% austenite and 50% ferrite. This improved both the strength and corrosion resistance. Duplex stainless steels have been studied and used more often in Europe than the United States. In Europe, they have been in production and use since 1930.

The Argon Oxygen Decarburization Process

The major problem in producing ferritic and austenitic stainless steels was the need for very low carbon content to avoid the precipitation of chromium carbides, which lowers the chromium content needed for corrosion resistance. This was especially a problem during welding, where a region in the heat-affected zone was at the right temperature for carbide formation.

Union Carbide Corporation (UCC) had done some work using an oxygen lance to reduce the carbon level. It was not successful due to the uncontrolled temperature of the bath. They hired a recent graduate from MIT named William A. Krivsky to work at the metals research laboratory to improve the oxygen process. Krivsky tried to add inert argon gas with oxygen to control the bath temperature and carbon reaction. He was successful in reducing the carbon to very low levels without excessive loss of chromium. With this laboratory success, UCC looked for a stainless producer to scale up the process to production levels. Only one company was interested, Joslyn Steel, a small producer in Fort Wayne, Indiana.

Over a period of years and many heats of steel, Joslyn could not produce a satisfactory result. During this time, they were introducing the oxygen and argon into the electric arc furnace with ceramic-coated lances. Joslyn and UCC finally realized that the refining was going to need a separate vessel in which the argon was blown through the bottom of the vessel but the oxygen was still introduced with a lance. The new vessel was similar to the old Bessemer converter. This became the successful argon oxygen decarburization (AOD) process. Union Carbide started offering licenses for the process in 1970. As industry began to see the advantages of the process, the number of vessels installed reached over 100 within a dozen years. Eventually, 75% of world production used the AOD process. It was the greatest technological advance in the history of processing stainless steel.

New Uses

A major use for stainless steel after WWII was for the exterior walls of high-rise buildings. The austenitic grade with 18% chromium and 8% nickel was used for its corrosion resistance, formability, and added strength when cold rolled. The first design was the Socony-Mobil Building on 42nd Street in New York City (**FIG. 7.9**). Within a few years, it was common practice to clad skyscrapers with stainless.

FIG. 7.9 The Socony-Mobil Building was the first high-rise building to be completely clad in stainless. Source: Photo by John W. Cahill, from The Skyscraper Center of the Council on Tall Buildings and Urban Habitat.

FIG. 7.10 Modernist architect Frank Gehry's Walt Disney Concert Hall in Los Angeles. Source: Carol M. Highsmith Archive, Library of Congress, Prints and Photographs Division.

The most spectacular building clad in stainless steel is the Walt Disney Concert Hall in Los Angeles, designed by architect Frank Gehry (**FIG. 7.10**). The Gateway Arch in Saint Louis, a U.S. National Historic Landmark, was built to commemorate the migration to settle the western United States. It was designed by Eero Saarinen and is the tallest stainless steel arch in the world at 630 feet. The arch is clad in over 800 tons of stainless steel (**FIG. 7.11**).

Home uses started with kitchen sinks in the late 1930s. Later, stainless steel expanded to the exteriors of refrigerators, stoves, dishwashers, and other appliances (**FIG. 7.12**). Stainless steels have largely disappeared from auto trim and hubcaps; the main automotive use is now exhaust systems with catalytic converters and fuel injection systems. The chemical, pharmaceutical, and electrical power industries use large amounts of stainless in piping, tanks, pumps, and other equipment.

Stainless steel has found many applications beyond its original uses in cutlery and tableware. It is the ideal metal alloy for designs requiring steel that does not rust.

FIG. 7.11 Built in the 1960s, the Gateway Arch welcomes over a million visitors per year. Source: STL_Skyline_2007_edit.jpg, Buphoff derivative work.

FIG. 7.12 A modern stainless steel kitchen. Source: Stainless Steel Kitchens Inc.

REFERENCES

1. C.A. Zapffe, *Stainless Steels: An Elementary Text for Consumers*, American Society for Metals, 1949
2. W. Giesen, The Special Steels in Theory and Practice, *Carnegie Scholarship Memoirs*, Vol 1, The Iron and Steel Institute, 1909
3. P. Monnartz, Iron-Chromium Alloys with Special Consideration of Resistance to Acids, *Metallurgie*, Vol 8, 1911
4. W. Borchers and P. Monnartz, German Patent No. 246,015, 1910
5. E. Maurer and B. Strauss, German Patent No. 304,126, 1912

Selected References

H.M. Cobb, *The History of Stainless Steel*, ASM International, 2010

L. Guillet, *Rev. Metall.*, Vol 2, 1905

E. Haynes, *Proc. Eng. Soc. Western Pennsylvania*, Vol 35, Engineers' Society of Western Pennsylvania, 1920

C.B. Horsley, The Midtown Book—The Chrysler Building, *The City Review*, http://www.thecityreview.com/chryslerb.html

J.W. Kerr, *Illustrated History of Budd Railway Passenger Cars*, Delta Publications, 1981

W.A. Krivsky, The Linde Argon-Oxygen Process for Stainless Steel, *Metall. Trans.*, Vol 4 (No. 6), 1994, p 1439–1447

M. Liljas, *80 Years with Duplex Steels: A Historic Review and Prospects for the Future*, Outokumpu Stainless AB, Sweden, 1994

J. Olsson and M. Liljas, 60 Years of Duplex Stainless Steel Applications, Paper 395, presented at Corrosion '94 (Baltimore, MD), National Association of Corrosion Engineers International, 1994

A.M. Portevin, *Comptes Rendus*, Vol 153, 1911

Chapter Eight

Aluminum—The Light Metal

Aluminum is a common structural metal in today's industrial world. It is produced in quantities exceeded only by steel. Its light weight, corrosion resistance, and ability to produce high strength with precipitation hardening heat treatment, and the general availability of bauxite ores around the world, make aluminum attractive for uses ranging from kitchen utensils and foil to airplanes and spacecraft.

Early attempts to extract aluminum from its ore by H.C. Oersted (1825) and Friedrich Wöhler (1827) had limited success. In 1845, Wöhler produced a sufficient amount to measure aluminum's specific gravity and ductility, and he determined that it melted at the low temperature of a blowpipe flame. This work alerted the scientific community to aluminum's light weight, approximately one-third that of steel.

The first minor production of aluminum was in 1854 by the Frenchman Henri Sainte-Claire Deville, who improved on Wöhler's process by substituting sodium for the more expensive potassium. He managed to produce marble-sized aluminum particles. Sainte-Claire Deville's work caught the attention of Napoleon III, who saw aluminum's potential as a material for military use. With government funding, Deville produced bars of aluminum that were exhibited at the Paris Exposition of 1855. This was France's first chance, after the Great Exhibition of 1851 in Britain, to show the world her technical capabilities, and the aluminum attracted special attention. Material from Deville's production was used to form tableware to replace gold and silver at royal banquets.

Within a few years, Deville reduced the price of aluminum from $90 a pound to $17 a pound. This was accomplished by using fluorspar and cryolite as fluxes. Sodium was a major cost in Deville's process, which required three pounds of sodium for each pound of aluminum produced. By 1885, discoveries in sodium

FIG. 8.1 The aluminum cap on the Washington Monument. Source: Wikimedia Commons/David Bjorgen.

processing reduced this cost to about one-fourth, and a British firm by the name of Aluminium Company Ltd. was formed to produce aluminum by Deville's process. The company produced 500 pounds a day by 1889, but was forced to close in 1891 after a total production of 250,000 pounds. The Deville process went out of use after 40 years of limited production. The earliest use of aluminum in the United States was for the cap on the Washington Monument (**FIG. 8.1**). The cap was a casting made from aluminum, refined and cast by the Frishmuth Foundry of Philadelphia. It received the ASM Historical Landmark designation in 1986.

The Invention

The first commercially successful process for producing aluminum was created by two young men living an ocean apart who invented the same new process

FIG. 8.2 Charles Martin Hall in 1905.

for reducing aluminum from its ore. Both men were just 22 years old at the time. One was an American chemist, Charles Martin Hall (**FIG. 8.2**), and the other was a French metallurgist, Paul Heroult. Their process would be the world's method for making aluminum for the next 125 years and beyond, and would propel aluminum and its alloys into use as the second major structural metal.

The driving force behind the search for low-cost aluminum was not the demand for the product, but rather the knowledge of the vast resource of aluminum-containing minerals within easy reach in the earth's crust. It had long been known that large deposits of aluminum-containing clays held great quantities of aluminum, especially in ores called bauxite, named after a region in France called Les Baux-de-Provence where the ore was first identified. Bauxite is scattered around the globe, mostly in tropical climates.

Charles Martin Hall

Charles Martin Hall was born in the village of Thompson in Geauga County, Ohio, in December 1863 (Fig. 8.2). The Hall family moved to Oberlin, Ohio, about 20 miles southwest of Cleveland, when the children were old enough for college. Even though Oberlin College was a liberal arts school, a number of science courses were taught, including chemistry, taught by Professor Frank Fanning Jewett, whose courses Hall took.

After graduating in the spring of 1885, Hall set up his now-famous experiments in a woodshed behind the family home with the assistance of his sister Julia, also an Oberlin chemistry student. During just a few weeks in the winter of 1886, he developed his method for making aluminum. His invention entailed dissolving aluminum oxide in the mineral cryolite. The high-temperature molten bath was contained in a pot lined with graphite, and graphite electrodes were inserted into the bath. A low voltage/high amperage electric current was passed from the electrodes to the lining, where aluminum metal was separated from the oxide. Aluminum collected at the bottom of the pot and was drained off periodically. Hall immediately applied for patent protection for his invention, and his patent was formally filed on July 9, 1886. However, the U.S. Patent Office notified him that an application for the same process had been filed on April 23, 1886 by Paul L.T. Heroult, a Frenchman. Heroult's patent application predated Hall's by two and a half months. The patent interference was resolved when Hall proved that his invention was actually in practice as of February 23, 1886, where Heroult had only his filing date of April 23, 1886. Hall, therefore, became the inventor of record in the United States by a mere two months.

Hall set out to obtain financial supporters to carry his process into production. His first successful contact was with the Cowles brothers who owned the Cowles Electric Smelting and Aluminum Company in Cleveland. The Cowles had a process for making an alloy of copper containing aluminum, which was called aluminum bronze. Hall joined the Cowles' company at their plant in Lockport, New York, with a salary of $75 a month for three months. If his experiments were satisfactory after that time, he would receive $750 and the company would continue to support his work. The option of further rewards for Hall was never fulfilled due to disagreements between Hall and the Cowles brothers, and Hall parted company with the Cowles organization before any progress was made in producing aluminum.

A Pilot Plant for Making Aluminum

However, Hall had worked with a manager at the Cowles plant, Romaine C. Cole, who had done some work on aluminum for a testing company in Pittsburgh named Hunt and Clapp. When it became clear to Hall that the Cowles brothers were not supporting him, Cole recommended that he and Hall contact Captain Alfred E. Hunt from Hunt and Clapp. Cole arranged a meeting in July 1888 to discuss the formation of a company to support Hall's experiments. The goal was to scale the process up to production. A second meeting between Hall, Cole, Hunt, and a small group of Pittsburgh men whom Hunt knew from the steel industry and his testing laboratory formed the technical and financial support of what would become the aluminum industry in America. Hunt and the financial

FIG. 8.3 Interior of the Smallman Street pilot plant of the Pittsburgh Reduction Company.

supporters agreed to organize a company called the Pittsburgh Reduction Company and to raise $20,000 to build and equip a pilot plant to demonstrate the value of Hall's process in producing low-cost aluminum (**FIG. 8.3**).

Next to Hall, Hunt was the most important man in developing aluminum as a commercial business (**FIG. 8.4**). While Hall and others were primarily interested in and worked on technology and production, Hunt carried the burden of financing, managing, marketing, and selling the product. Alfred Ephraim Hunt was born in Massachusetts in 1855. He graduated from MIT in mining and metallurgy in 1876. In 1887, he and George Clapp bought out the owners of the Pittsburgh Testing Laboratory and renamed it Hunt and Clapp.

The Pilot Plant

Hall and Cole built a pilot plant on Smallman Street in Pittsburgh and set up reduction cells and a motor-generator set to provide electricity for the operation. Their first production, after numerous start-up problems, yielded a few marble-sized particles on Thanksgiving Day 1888 (**FIG. 8.5**). Thus, the modern world of aluminum was born.

Hall and Cole were not compatible as fellow workers in running the pilot plant, so Cole was replaced by a young worker from Hunt's testing laboratory,

FIG. 8.4 Alfred E. Hunt. Source: www.webcemeteries.com.

FIG. 8.5 Aluminum nodules produced by Charles Martin Hall in 1886 using an electrolytic process.

FIG. 8.6 Arthur Vining Davis replaced Romaine C. Cole in operating the aluminum pilot plant. Source: www.alcoa.com.

Arthur Vining Davis (**FIG. 8.6**). Hall and Davis increased production at the pilot plant to 50 pounds per day, which sold for $8 a pound. By adding two new dynamos, production increased to 450 pounds per day, and they decreased the price to $2 per pound. Realizing the need to achieve higher production levels to obtain economies of scale, and in need of funds to meet daily obligations, the company's principals approached the Mellon Bank for a $4000 loan. The Mellon brothers, Andrew and Robert, supplied a larger loan, $25,000, and later took a position in the company by buying 60 shares from Hall at $100 per share.

A New Production Plant

The Pittsburgh Reduction Company outgrew the pilot plant on Smallman Street. The Mellon brothers encouraged them to locate a production plant in New Kensington, northeast of Pittsburgh. By now, Hall had discovered that, by increasing the size of the reduction pots and therefore the electric current, he could maintain the heat, keeping the molten bath at the necessary temperature without an outside heat source. Also, by scaling up the operation, less energy was needed per pound of aluminum produced. The Mellons purchased a stock offering at this time of 500 shares at $60 per share. This allowed them seats on the board of directors. The Mellons were now insiders and not just bankers providing loans.

The Move to Niagara Falls, New York

In 1893, the Pittsburgh Reduction Company signed the first industrial contract to take electricity from hydropower produced at Niagara Falls, New York. A new processing plant built there became operational in 1895. A second plant was constructed at the Falls in 1896, followed by a third plant, with its own electric power facility, built in 1906. With its steadily increasing capacity to produce aluminum and the decreasing cost of production, the Pittsburgh Reduction Company expanded rapidly throughout the rest of the 1890s. Production reached one million pounds in 1896, 2.4 million pounds in 1897, and five million pounds in 1900. The price decreased during this same period from $0.78/lb for ingot metal in 1893 to $0.48 in 1896, $0.36 in 1897, and finally to $0.33 in 1900. However, this still left aluminum in competition with steel selling at several cents a pound. Even aluminum's weight advantage of 3 to 1 compared to steel could not overcome such disparity in price. Also, aluminum was producing five million lbs/year in 1900, or 2500 tons; ten million tons of steel were produced that year. In fact, the steel industry consumed a large portion of the total aluminum produced. Aluminum added to molten steel removed harmful gaseous oxygen by forming aluminum oxide.

Early Aluminum Applications

Production of aluminum increased from five million pounds in 1900 to 35 million pounds in 1909. A major application was kitchenware. The Pittsburgh Reduction Company acquired a kitchenware producer in 1901 when the producer went bankrupt due to owing a substantial bill for aluminum. This acquisition brought into the company a pair of college students who had been selling pots and pans door to door with great success. A new company was organized to improve the quality and expand this effort. The resulting kitchenware became well known under the WearEver Cookware brand and was widely sold as high-quality utensils (**FIG. 8.7**). Wire for electrical conductors was another growing product, and later, with the development of a system of stranded cable that included a steel wire for strength, aluminum was able to compete with copper for long-distance electrical transmission of high-voltage alternating current.

The need for wire, sheet, plate, and other product forms meant that the company had to find other mills willing to convert ingot to these products or develop the capability themselves. At this early stage in the aluminum business, other metals producers, mainly steel and copper producers, were not interested

FIG. 8.7 The Pittsburgh Reduction Company began producing WearEver aluminum cookware in 1902. Source: www.wearever.com.

in helping a new competitor get established. In addition, the rolling and working of aluminum presented problems that other mills were not interested in solving. Therefore, in the earliest days at the New Kensington plant, Hall and Davis installed equipment and hired specialists to fabricate wire and sheet. When all the aluminum chemical reduction operations were moved to Niagara Falls under Hall, the working of aluminum into other products was concentrated at New Kensington under Davis.

The first decade of operations saw small annual losses and modest profits until the end of the decade; pretax earnings reached $181,000 in 1899 and $322,000 in 1900. By that point, the company had declared total dividends of $300,000 and had total retained earnings of $636,000. All of this had been accomplished despite a severe recession between 1893 and 1900. Undoubtedly, having the Mellons as partners and bankers was very helpful during those difficult times. At this critical time in the aluminum business, the Pittsburgh Reduction Company lost its dynamic leader, Alfred E. Hunt. With his enthusiasm for the military, he had led his company of men into the Spanish American War. He was posted to Puerto Rico, where he contracted a tropical illness that took his life shortly after his return to the United States.

Richard Mellon was appointed president to succeed Hunt, but the day-to-day management of the company was under Arthur Vining Davis. This did not represent a significant change because Davis had been supervising the manufacturing operations at New Kensington, with aluminum production under Hall at Niagara Falls. The company continued to be controlled by Hall, Davis, the Mellons, and the small group of original investors. The family of Alfred E. Hunt inherited his shares in the company, and his son Roy Hunt came into the company as an employee after his graduation from college. Roy Hunt became a key manager under Davis and played a significant role in the future of the company.

After 1900, a growing demand for aluminum came from the developing automobile industry. It was easier to manufacture custom bodies from the more ductile aluminum rather than steel (**FIG. 8.8**). The company became a supplier of fenders and bodies in the first decade of the 20th century. This application, as with so many others, was short-lived, as mass production of cars required the lowest-cost material, steel. A market for aluminum cast parts was a substantial part of this early automobile industry. They were used for transmission cases, rear axle housings, and other parts. A new casting process where the

FIG. 8.8 Early auto manufacturers used aluminum for limited production, as seen in this 1924 Studebaker Light Six that features an all-aluminum body.

molten aluminum was poured into water-cooled steel molds rather than into sand molds opened the market for aluminum pistons to replace the much heavier cast iron. One aluminum automobile application, the piston, survived into modern times. Aluminum again looked attractive for auto parts after the oil shortages of the 1970s, when the government required improved fuel economy.

Raw Materials

The expanded production of aluminum metal after 1900 required the Pittsburgh Reduction Company to seek lower-cost raw materials and electrical power. In addition, this search for backward integration was a necessary part of the company plan to exclude competition after the Hall patents expired in 1909. Electrical power was the costliest ingredient in aluminum production. By 1909, when the patents expired, the company (renamed the Aluminum Company of America, or Alcoa for short) secured water rights along the Little Tennessee River. They built dams, power plants, and a smelter around a small settlement they called Alcoa, Tennessee.

Alcoa purchased the partially completed Southern Aluminum Company located in North Carolina in 1915. This company was the only competitor for Alcoa until after World War II. It was a French company that attempted to break through the high tariffs imposed on imports by the United States. World War I commenced before they completed construction and funds were not available to finish the plant.

Bauxite mining began in the United States in Rome County, Georgia, as early as 1883. Later, the mineral was found in greater quantities in Arkansas, where mining began in 1899. The Pittsburgh Reduction Company's first entry into raw material production was with the General Bauxite Company. In the last year of his life, Hunt traveled to Arkansas to purchase bauxite property. By 1909, the company acquired the General Mining Company of Arkansas and the Republic Mining and Manufacturing Company. With Arkansas bauxite as a secure source, the company built a large refinery to reduce the ore to alumina in East St. Louis, Illinois. From there, the alumina was shipped to Niagara Falls; Massena, New York; and the Canadian plants for processing into aluminum ingots. Because known reserves of bauxite in the United States were inadequate, Alcoa searched overseas for future supplies. Their largest investments were in the British and Dutch Guineas, where vast reserves were located.

The company began producing a synthetic cryolite to replace the natural mineral that was used to dissolve the alumina for the electrolytic processing to aluminum metal. The only available source of the natural cryolite was in Greenland, where it was mined and sold exclusively to the Pennsylvania Salt Company Mine in the United States.

World War I

By the start of World War I, Alcoa had fully integrated the aluminum production process, from mining to aluminum metal production. They had also moved forward into production of many finished products. Alcoa became a major force in the worldwide aluminum industry. There were a number of overseas competitors, especially in France and Switzerland, but Alcoa was protected by high tariffs during the prewar years and by the wartime needs after 1915. The company was able to sell its inventory of aluminum to the European powers allied against Germany in 1915 and 1916. Production was then purchased for the United States' defense use in the remaining years of the war. The bulk of all of the aluminum sold in this period went into munitions. As a powder, aluminum was mixed with ammonium nitrate to form a high-powered explosive. Many other defense applications helped promote postwar uses for the metal. The most prominent of these was the selection of aluminum for the Liberty engine used in most American aircraft built during the war (**FIG. 8.9**).

Alcoa had pretax earnings of $4 million to $6 million each year between 1909, when the patents expired, and 1914. These earnings increased to $9 million in 1915; they leaped to $25 million in 1916, $20 million in 1917, and $15 million in 1918. By 1919, Alcoa had corporate equity, mainly from retained earnings, of

FIG. 8.9 The Liberty L-12 aircraft engine used cast aluminum for light weight.

$100 million. However, $63 million of this equity was earned during World War I. Alcoa's strong position in later years was a direct result of the enormous profits from the war.

Ongoing Research

Numerous technical problems arose during the war years for which Alcoa was unprepared. Castings for the Liberty aircraft motor, engine pistons, and many other applications posed special problems for the aluminum foundry industry. The primary embarrassment for the company was their inability to replicate the new German alloy called Duralumin. This alloy was used for the 100 Zeppelins built during the war, and it was being made and used by France and Great Britain in limited aircraft applications. The delay in undertaking research and development was due to Hall's reluctance to employ personnel trained in science. Science was his field of expertise, and he jealously guarded it from others, especially outsiders. With his death in 1914, management moved to remedy this. At the same time, Zay Jeffries, trained in engineering and a self-educated metallurgist, began studying the aluminum casting problems at the Aluminum Casting Company in Cleveland. Zay Jeffries became a major expert in the metallurgy of aluminum (**FIG. 8.10**).

FIG. 8.10 Zay Jeffries was the first metallurgist to use the microscope to study problems in aluminum castings.

FIG. 8.11 Francis C. Frary, Alcoa's first director of research. Source: www.electrochem.org.

Alcoa hired a director of research in 1919. His name was Francis C. Frary, and he was a chemist from the University of Minnesota (**FIG. 8.11**). He oversaw Alcoa research for the next 33 years and was highly regarded throughout American industry. In 1946, he received the Perkins Medal from the American Chemical Society, which Hall had received in 1912. As corporate director of research, Frary took control of the laboratory in Cleveland in 1920, when Alcoa acquired the Aluminum Casting Company as payment for the debt owed on aluminum purchases made during the war. Thus, Jeffries and his coworkers came into Alcoa.

Alcoa built two teams of researchers during the 1920s, the one they inherited in Cleveland and another that Frary started in New Kensington, Pennsylvania. Jeffries continued as director of the Cleveland laboratory until 1923, when he became a consultant. During this time, he and Robert Archer, along with other researchers hired by Jeffries, developed numerous alloys. One of the first alloys developed by Jeffries and Archer was a non-heat-treatable casting alloy containing 5% silicon. This alloy is still used in many applications for its ease in processing. Another version of the 5% silicon alloy that could be heat treated became alloys 355 and 356. They, too, are in major use at the present time. Jeffries and Archer developed the earliest American heat treatable casting alloy (195) with additions of copper to aluminum. This alloy could be strengthened by the new precipitation hardening treatment discovered in Germany by Alfred Wilm. Other heat treatable alloys developed during this period included a forging alloy (25S) that was used in aircraft propellers and an all-purpose aircraft structural alloy (14S). They produced the first American alloy containing magnesium and silicon for age hardening (61S) and a new cast-piston alloy for automotive and aircraft engines (132) that is still used in several modified forms.

Other developments by researchers at Alcoa during the 1920s included a method, developed by Edgar H. Dix, for protecting aluminum sheet alloys from corrosion, especially from seawater. Dix's solution, which continues to be the major method in use today, bonds surface layers of a more corrosion-resistant unalloyed aluminum to both sides of the alloyed aluminum sheet. The highly alloyed heat treatable alloys are usually more prone to corrosion and corrosion cracking under stress than the unalloyed material. The penalty for this improvement in corrosion resistance is a reduction in strength due to the lower strength of the clad surface. In recent years, special alloys have been developed for cladding the higher-strength aerospace alloys.

Jeffries and his team of researchers at the Cleveland laboratory performed valuable work throughout the 1920s that strengthened Alcoa's position as the only American primary producer of aluminum. By 1930, when Alcoa published a technical book on the properties and processes of their aluminum products, the list of authors contained 25 employees in research, product development, and engineering. Many of these individuals continued to serve Alcoa in technical capacities through the 1930s and 1940s.

Alloy Development

With two laboratories working on research and development through the 1920s and 1930s, over 20 new alloys were added to the product line. The first new sheet alloy, called 17S, was based on the German Duralumin. It was successfully produced in the 1920s and was used to build the first commercial all-metal passenger aircraft in the United States, the Ford Trimotor (**FIG. 8.12**). About 200 Ford Trimotor planes were built in the late 1920s and early 1930s, but production ceased as the Great Depression deepened and the aircraft design, which carried 10 to 12 passengers, became obsolete for profitable commercial service.

Alcoa developed a new, higher-strength alloy called 24S in the 1930s. The major change from 17S to 24S was increasing the magnesium, from 0.5 to 1.5%. This increased the design strength of 24S from 40,000 to 50,000 psi. In addition, it was found that moderate cold working, such as stretching or rolling sheet material immediately after water quenching, followed by aging could further increase the design strength of 24S to 57,000 psi. These properties replicated those produced with Wilm's original room-temperature aging treatment, called natural aging. This alloy was the construction material for the first commercially successful passenger plane, the Douglas DC-3, in 1935 (**FIG. 8.13**).

FIG. 8.12 The Ford Trimotor was the first commercial aircraft built in the United States with 17S, a high strength aluminum alloy. Source: www.goldenwingsmuseum.com.

FIG. 8.13 The Douglas DC-3, the first successful passenger plane, was built with Alcoa's new 24S alloy.

Another age-hardening alloy system developed by Alcoa in the 1930s was based on adding 1% magnesium, 0.6% silicon, and 0.3% copper to aluminum. This alloy, called 61S, is still used today (now called 6061) and is the structural material for a great number of ordinary engineering applications. There are a number of alloys based on 61S that contain additional alloying elements. These alloys are outstanding for their ease of fabrication, corrosion resistance, and low cost compared to high-strength aircraft alloys. They feature design strengths of 35,000 to 50,000 psi and have excellent characteristics for general industrial applications, such as trucks, buses, rail cars, trailer tanks, storage tanks, building construction, and light aircraft. The large number of mill products for these alloys includes sheet metals, forgings, extrusions, bars, tubing, pipes, and wires.

Aluminum for World War II

Alloy 24S was the aluminum used for nearly all of the 300,000 planes built in the United States during World War II (**FIG. 8.14**). The quantity of aluminum needed for this vast undertaking greatly exceeded the capacity of Alcoa, the

FIG. 8.14 The Boeing B-17 Flying Fortress and nearly all of the 300,000 aircraft built during World War II used Alcoa's 24S alloy.

only aluminum producer in the country. In 1939, the United States produced 148,000 tons of aluminum, compared with 200,000 tons in Germany. Preparing for war drove the German demand. By 1943, when wartime production reached its peak, the United States produced 835,000 tons against 250,000 tons in Germany. In addition, Canada increased production from 75,000 tons in 1939 to 450,000 tons in 1943. Overall, production for the five war years reached 4,000,000 tons for the United States and Canada, versus 1,400,000 tons for Germany.

To meet the increased demand, numerous aluminum plants, financed by the U.S. government, were built and operated by Alcoa. These plants included four to produce aluminum oxide, eight to reduce the oxide to metal, and 10 to manufacture the finished product. These government-financed plants produced twice the aluminum of the Alcoa-owned plants. At the end of the war, the government plants were sold to Reynolds Metals Company and Kaiser Aluminum and Chemical Company, thus ending the Alcoa monopoly that had existed since 1888. With this new aluminum industry, the old system of identifying alloys was modified. The new system used the 2000 series for aluminum-magnesium alloys. Thus, 24S became 2024, which remains a major high-strength alloy, even though stronger alloys have been developed for the most critical applications.

In 1953, Alcoa built a new corporate headquarters to house its widespread activities (**FIG. 8.15**). Located in downtown Pittsburgh, it was the first aluminum-clad high-rise building.

Aerospace Applications for Aluminum Alloys

Many aluminum alloys were used in the hardware for the NASA space program. They made possible the lightweight design for the Destiny research laboratory, manufactured by the Boeing Company (**FIG. 8.16**). Destiny was launched in February 2001 and is attached to the larger space laboratory. The module is a cylindrical shell, 14.5 ft in diameter and 27.5 ft long. A debris shield surrounds the module to protect it from microparticles coming from outer space. Destiny is constructed from precipitation-hardened aluminum alloys.

The shell, its rings, and its outer case are made of 2219 alloy. This alloy was selected for its ease of welding and proven performance, as it was used in the Saturn V launch rocket. Alloy 2219 contains 6.3% copper and minor amounts of titanium and manganese. It was developed in the early 1950s and is a high-copper modification of an earlier alloy (2025) invented by Jeffries and Archer in 1921. The Destiny research lab access hatches are made of the higher-strength aluminum-zinc alloy 7075. They are made from plate and are 4 ft in diameter. After forming and machining, they are solution heat treated, aged to beyond peak strength for improved corrosion resistance, and plastically deformed in compression for stress relief.

FIG. 8.15 Built in 1953, the first aluminum-clad high-rise building served as Alcoa's corporate headquarters in Pittsburgh.

FIG. 8.16 Destiny, NASA's laboratory module for U.S. research payloads aboard the International Space Station, was built by the Boeing Company. Source: NASA.

Recent New Alloys

Aluminum alloys containing additions of zinc along with magnesium and copper were originally studied in Germany. These alloys with zinc as a major alloying element produced very high strengths but were prone to crack under stress when exposed to corrosion. Nevertheless, research on these alloys was performed at Alcoa, and the first composition used commercially was for aircraft propellers in 1940. Later, the problem of corrosion cracking under stress was greatly reduced by adding small amounts of chromium. This led to the commercial alloy 75S (now 7075), which contained 5.5% zinc; it was introduced during World War II as the structural metal on the B-29 long-range bomber. This alloy was artificially aged to design strengths of 73,000 psi. A modification of this alloy with 6.8% zinc (7178), introduced in 1951, developed strengths as high as 78,000 psi.

A recent alloy developed by Alcoa, called 7055, contains 8.0% zinc, 2.3% copper, and 2.0% magnesium. It was designed to provide exceptionally high strength in compression. Strength levels of 90,000 psi are achieved in plate and 97,000 psi in extrusions. The strength is 10% higher than the best previous alloy and 25% higher than the original aluminum-zinc alloy 7075, developed during World War II. Alcoa received the ASM International Engineering Materials Achievement Award for the development of 7055.

These latest high-strength alloys from Alcoa are used on the Boeing 777 (**FIG. 8.17**). The major structure of the plane is constructed with two basic

FIG. 8.17 A new high-strength alloy, 7055, is one of several alloys used on the Boeing 777 all-aluminum aircraft. Source: Wikimedia Commons/United Airlines.

aluminum alloys. The upper wings, because they are the most highly stressed components, are built of 7055 alloy. The lower wings are constructed of modified versions of 2024, specifically 2224 and 2324. They have lower impurity content, which improves toughness. The fuselage is made of 2524, another version of 2024 that contains even lower levels of impurities for greater toughness.

Aluminum has come a long way since the early days of pots and pans. With the discovery of age hardening as a mechanism for increasing its strength, aluminum has a major place in modern technology as a structural metal with unlimited uses.

Selected References

D. Donaldson and B. Raahauge, Ed., *Essential Readings in Light Metals, Alumina and Bauxite*, John Wiley & Sons, 2013

D.R. Duggan and G. Williams, *Alcoa*, Arcadia Publishing, 2011

S.L. Hoyt, *Men of Metals: An Exciting Career among the Trailblazers of Modern Metallurgy*, American Society for Metals, 1979

I. McNeil, *An Encyclopedia of the History of Technology*, Routledge, 2002

W.D. Mogerman, *Zay Jeffries*, American Society for Metals, 1973

G.D. Smith, *From Monopoly to Competition: The Transformation of Alcoa, 1888–1986*, Cambridge University Press, Cambridge, U.K., 1988

R.F. Tylecote, *A History of Metallurgy*, 2nd ed., Maney Publishing, 1976

R.M. Young, *Made of Aluminum: A Life of Charles Martin Hall*, David McKay Company, Inc., 1965

Chapter Nine

Titanium—A New Metal for the Aerospace Age

William Gregor, an English clergyman and amateur chemist, discovered titanium in 1790. It was rediscovered in 1795 by an Austrian chemist, Martin Heinrich Klaproth, when studying the mineral rutile. It was Klaproth who named the new metal titanium after the Titans, the giant sons of the earth. Titanium is element No. 22 in the periodic table. Its density is 4.5 g/cm^3, midway between aluminum at 2.7 and iron at 7.86. It has a melting point of 1812 °C (3290 °F), compared with 1535 °C (2785 °F) for iron. The low density and high melting point compared with iron indicated a metal with structural potential. When its resistance to corrosion was discovered, this further enhanced the uses for which titanium was uniquely suited. It seemed to be the ultimate answer for a strong, lightweight, corrosion-resistant metal for numerous structural designs.

The advantages of titanium looked promising in the late 1940s and early 1950s to a military on the cutting edge of developing jet engines, supersonic aircraft, missiles, and lighter-weight trucks, tanks, landing craft, and other hardware for the Cold War and the Korean conflict. The problem was the lack of industry producing titanium.

Titanium Made at General Electric Corporation

The first attempt to make titanium metal in the United States was at the General Electric Research Laboratory in Schenectady, New York. General Electric was looking for a light bulb filament to replace the cotton originally used by Thomas Edison. Tungsten was thought to be the likely candidate, but a process was not yet available to produce ductile tungsten. Titanium was believed to have a high melting point, which was a requirement for filaments. A young graduate from New Zealand with an advanced education in Europe was on his way home in 1905 when he stopped in the United States and was hired by General Electric to explore the possibility of finding a process to produce titanium metal. His name was Matthew Hunter, and he had enough success to be regarded as the first research engineer to produce a small quantity of titanium metal (**FIG. 9.1**). He determined that the melting point of titanium was about 1800 to 1850 °C (3270 to 3360 °F), not 6000 °C (10,830 °F) as previously thought. At this point, General Electric lost interest and Hunter moved on to nearby Troy, New York, to become a professor at Rensselaer Polytechnic Institute (RPI), where he spent the rest of his career. He was considered the pioneer of titanium in later years when it became a common metal in the aerospace age.

Another early attempt to make titanium was in the 1920s at the Philips Company in Eindhoven, the Netherlands. This process involved the thermal dissociation of titanium iodide and the deposition of the titanium on a hot tungsten wire. Many years later, samples of this material were studied at the United States Bureau of Mines (USBM), which generated interest in pursuing the development of a more practical process for titanium production.

FIG. 9.1 Matthew Hunter produced small amounts of titanium for laboratory use. Source: Wikimedia Commons.

Invention of a Process for Making Titanium

In May 1940, a middle-aged research engineer immigrated to the United States from Luxembourg with a process and a patent for making titanium. He fled Luxembourg just before the German invasion. His name was William J. Kroll, and his titanium process became the basis for the titanium industry for the United States and overseas (**FIG. 9.2**). This was not the first time that Kroll had been to the United States. He was in the United States in 1932 during an attempt to sell the value of titanium to American industry, but he failed to stimulate interest at that time. Until he discovered his titanium process, only laboratory quantities of titanium had been produced, and the limited work performed on these samples had not led to further effort.

Kroll was born in 1889 at Esch, Luxembourg. He attended a Jesuit high school, founded in 1603, where he received an education in the classics and in chemistry and physics. He earned his doctor of engineering in metallurgy from the Technische Hochschule in Charlottenburg, Germany in 1917.

After several years of employment in Germany, Austria, and Hungary, Kroll established his own research laboratory in a large home he purchased in his native Luxembourg (**FIG. 9.3**). There he conducted research in metallurgy and electrochemistry. He worked as a lone experimenter with only one secretary and a laboratory assistant. He never married; his work became his life. His research resulted in development of aluminum, magnesium, copper, and lead

FIG. 9.2 William Kroll, inventor of the process for making titanium metal. Source: www.france-metallurgie.com.

FIG. 9.3 William Kroll's home and laboratory from 1923 to 1940. Source: www.wort.lu/Gyorgy Foldes.

alloys. His initial interest in titanium stemmed from his attempts to substitute beryllium in the high-strength copper-beryllium alloy in 1928.

Kroll had some financial help for research from the German firm of Siemens & Halske. When they lost interest in supporting his work, Kroll obtained control of the foreign patent rights and invested his own funds to continue development in his private laboratory. By 1938, he had produced 50 pounds of titanium metal. He made a second trip to the United States to again generate interest in his work. He visited major American corporations including General Electric, Westinghouse Electric, International Nickel, and Union Carbide. He also visited a former classmate from the Technische Hochschule, an American named Samuel Hoyt, who was a metallurgical manager at A.O. Smith. Once again, he failed to find an interest in his process or in titanium.

The first published report on Kroll's titanium process was a paper he presented at the 1940 meeting of the Electrochemical Society. That same year, he was issued United States Patent #2,205,854 for his invention. Kroll's titanium process was a modification of the sodium reduction of titanium tetrachloride, used by Hunter. Kroll used magnesium instead of sodium as the reducing agent.

FIG. 9.4 Titanium sponge after cleaning and before compacting for melting. Source: Wikimedia Commons.

The resulting metallic product was a sponge-like collection of particles that had to be cleaned by acid leaching of residual magnesium and magnesium chloride contaminants (**FIG. 9.4**). The sponge was then crushed to powder, compacted, and heated at a high temperature to sinter it into a solid mass. The resulting metal generally contained impurities, which affected its properties.

The USBM Metallurgy Division, headed by Reginald Dean, took an interest in titanium metal based on limited property measurements of laboratory material made at the Philips Company in Europe. The effort was initiated in 1938 but made little progress before the beginning of World War II. The Bureau of Mines began experimenting with Kroll's process under the direction of Frank S. Wartman at the Utah Engineering Experiment Station during World War II. Early researchers included James R. Long, O.C. Ralston, and Frank Cservenyak. Titanium was being made in 15-pound batches by 1944.

At this stage of development, Kroll became a consultant to the Bureau of Mines at Albany, Oregon. His process was being used to develop a new metal, zirconium, for use in atomic reactors. Zirconium is in the same group in the periodic table as titanium and responded to Kroll's process. Although zirconium was heavier than titanium, it had a low absorption rate for neutrons, and thus it was selected for atomic reactors for the U.S. Navy's nuclear submarine program.

Kroll's process and the support needed to carry on the research on titanium came together at a new Bureau of Mines (BOM) plant in Boulder City, Nevada. The only thing missing for Kroll was his patent, which the Office of Alien Property Custodian had confiscated during the war. Kroll had to fight through the courts to regain custody.

Studies of Properties

Titanium made its first advance beyond the USBM at Boulder City, Nevada, in 1946, when a contract was awarded to Battelle Memorial Institute in Columbus, Ohio, to study the properties, welding, and fabrication characteristics of titanium and its alloys. The USBM in Boulder City supplied the titanium, and funds were supplied by the Air Force and the Douglas Aircraft Company.

Battelle Memorial Institute was founded in 1929 with an endowment from the will of Gordon Battelle. Later, an additional endowment came from his mother. The Battelle family fortune was made in the iron and steel industry in southern Ohio. Gordon Battelle believed that the steel industry would benefit technically and economically from modern research. Battelle Memorial Institute was a contract research operation; it performed research for industry and government agencies on a fixed-fee basis. Battelle supplied the research staff and laboratory equipment, and the contractor paid for services. In this manner, companies, associations, and the government who did not have the appropriate staff or laboratory facilities could contract for research studies.

Every branch of the armed forces, as well as all the government agencies with an interest in this new metal, followed the research work at Battelle via progress reports. The research work became the first independent study of broad scope to evaluate the practical capabilities of titanium. After a two-year study, a final classified report was issued in 1948. A major conclusion was that titanium and titanium-base alloys had great potential engineering importance and further development should be diligently pursued. This report and other evidence from work at the BOM in Boulder City, Wright-Patterson Air Force Laboratories, and the Army Ordnance Department at Watertown Arsenal provided the foundation for promoting a new industry based on titanium.

The BOM in Boulder City needed financial help in 1947 to continue pilot plant production. It was supplied by a contract from the Navy for $25,000. The product was evaluated at P.R. Mallory and Co. of Indianapolis by pressing and sintering powder. One of the research engineers, L.S. Busch, and a manager, Frank F. Vanderburgh, later made significant contributions at a titanium supplier named Mallory-Sharon Titanium Corp.

Sponge Production

E.I. du Pont de Nemours and Company (DuPont) was the first commercial organization to commit resources to producing titanium. In late 1948, DuPont offered titanium sponge for sale at $5 per pound and they produced 100 pounds per day. DuPont was a large producer of titanium dioxide, which was used in paint and many other products as a whitener. Thus, it was natural for them to be the first commercial producer of titanium sponge made by Kroll's process from titanium tetrachloride ($TiCl_4$). The Remington Arms Division of DuPont did research on the metal made from this sponge, and later became a major partner in producing metal for the aerospace market.

The first major conference on titanium was held in Washington on December 16, 1948. It was organized by Nathan Promisel of the Navy's Bureau of Aeronautics and Julius Harwood of the Office of Naval Research, and was held at the National Academy of Sciences. An assembly of 200 attendees heard papers presented by authors from Battelle, Remington Arms Division of DuPont, P.R. Mallory, and the Navy's Bureau of Aeronautics. This meeting promoted great enthusiasm for the need for titanium and was considered the launch of this new metal. An endeavor starting in 1950 required a decade to accomplish, utilizing dozens of metallurgical engineers in industry, universities, and other research laboratories, and several hundred million dollars of investment by the federal government and the industries that became involved in production.

Many factors came together at this time that influenced all levels of government to cooperate. All through World War II, there were shortages of certain materials vital to the war effort. Metals such as tungsten, nickel, chromium, manganese, and others were controlled by the War Production Board and released on a priority basis for only the most critical applications. The initiation of Korean hostilities in June 1950 provided additional incentive for developing a new metal with promising properties and ready availability in the United States. It was believed at this time that the United States was self-sufficient in rutile, the major ore of titanium. The Cold War was underway, and the United States was supplying Berlin with all its vital supplies by aircraft, which showed the need for lightweight equipment for airborne use.

The development of a new metal was a daunting task. As an example, it took 30 to 40 years to bring aluminum and stainless steel from their original inventions to full commercial practice. For titanium, there were many pitfalls along the way. Uses were attempted as fads due to the publicity given the new material. Unanticipated problems arose during early stages, including catastrophic failures endangering lives and property. Many technical difficulties arose regarding fabrication, welding, corrosion, and inappropriate applications. And then there was the struggle to develop alloys, as there were few applications for unalloyed titanium. It generally lacked the strength for structural designs. The search for alloy additions and the necessary heat treatment to develop the properties for the high strength needed in industry required a large amount of research and development. Numerous alloys were usually needed to fill the various requirements.

When sample quantities of the BOM Boulder City material became available, they were distributed to interested laboratories across the country, including those of the Army, Navy, and Air Force, as well as those of some industrial companies, such as National Lead Co. and Kennecott Copper Corp. Studies of the material revealed properties of immediate interest, especially corrosion resistance and strength. The metal gained numerous advocates within the military and industry. Early promoters visualized the use of titanium in naval vessels, armor plate, tanks, trucks, landing craft, aircraft structures, and airborne equipment. It was thought that it could possibly replace both aluminum and steel in the design of many defense applications. Industrial uses would be based on its combination of properties, featuring light weight, corrosion resistance, and high strength.

Enthusiasm for titanium reached the highest levels of the Defense Department, where funding was obtained for further research and development. One of the earliest studies in search of alloys with superior mechanical properties was the RAND Project of the Air Force, Air Materiel Command. This program financed a laboratory project at Battelle Memorial Institute with the objective to screen a wide variety of alloy additions for their effect on both the physical and mechanical behavior of titanium.

The first goal of the Battelle project was to design a furnace that could melt titanium powder or sponge in an atmosphere free of oxygen, nitrogen, and hydrogen. It was known that these gases, which are present in air, were harmful to titanium metal. This was accomplished by constructing a water-cooled copper crucible, in which a vacuum was maintained as the powder or sponge was fed into a very high temperature electric arc. In this manner, numerous small melts were made, containing various quantities of assorted alloy additions. The experimental examination of these small ingots revealed hardness and strength, and the physical effects of the alloy content improved the behavior of titanium when heated to high temperatures, either for hot working to shape it or for heat treating it to improve mechanical properties.

Some metals, such as iron and titanium, undergo a change in crystal form upon heating to a high temperature. This is called allotropic transformation. Above 882 °C (1620 °F), titanium has a body-centered-cubic (bcc) structure. Upon cooling below that temperature, the atoms rearrange into the hexagonal-close-packed (hcp) structure. Alloy additions change the transformation point from a single temperature to a range of temperatures, so that there is an area in the phase diagram where the two crystal structures coexist. The exact temperature range where the phase boundaries exist is very important in processing and heat treating alloys. To gain such information about titanium, with the goal of developing titanium alloys of high strength, the defense agencies funded the measurement of many phase diagrams.

The first program, funded by the Air Force Air Materiel Command, conducted a project covering a large number of alloy additions to titanium. The published papers in the *Transactions of the American Institute of Mining and Metallurgical Engineers* in 1950 showed data on hardness, strength properties, and phase diagram studies.

At about the same time that the Air Force work was underway, Kennecott Copper Corp. contracted a program at the Armour Research Foundation (ARF) in Chicago. ARF was formed in the late 1930s to perform research for industry, similar to Battelle. The ARF, however, was affiliated with the Illinois Institute of Technology (IIT) in Chicago and did not conduct research in metallurgy until after World War II.

While limited research activity was underway, several steel producers began to show interest in titanium. They were the stainless steel producers Allegheny Ludlum Steel Corporation, Crucible Steel Company, and Republic Steel Corporation and a specialty steel producer, Sharon Steel Corporation. These companies were attracted to titanium for its interesting properties. They thought they might be able to exploit titanium as a new business. They were also keeping an eye on titanium because they thought it might become a serious competitor to their regular business, stainless steel. During 1949 and into 1950, these companies bought titanium sponge from DuPont, then melted and fabricated it into bar and sheet product. The material was used primarily for testing by various defense laboratories, universities, and industrial firms.

The National Lead Co. (NL) came on the scene in 1950 as a second producer of sponge. They were doing pilot plant work at Sayreville, New Jersey, and Niagara Falls, New York. The biggest producer of titanium dioxide was NL. They took over production of sponge under contract with the BOM Boulder City Experimental Station.

The year 1950 was of major importance in forming the new titanium industry. A joint venture was formed between Allegheny Ludlum Steel Corporation and NL to produce sponge, melt and cast the resulting metal into ingots, and supply metal products. This entity was called Titanium Metals Corporation of America (TMCA, later TiMet).

A second arrangement was made between the Remington Arms Division of DuPont and the Crucible Steel Company, called Rem-Cru Titanium Inc. The following year, a third combination was formed between P.R. Mallory & Co. and Sharon Steel Corporation called Mallory-Sharon Titanium Corporation.

This latter group did not have a sponge facility but bought their sponge from DuPont.

Research activities accelerated in 1950 and 1951. The Army Ordnance Corps at Watertown Arsenal began a large in-house research initiative and a major program of contract research with outside firms. They funded alloy development at Allegheny Ludlum, a variety of programs at Battelle and ARF, and phase diagram studies under Professor John Nielson at New York University (NYU). Nielson built and staffed a new research division at NYU to conduct work on titanium. This group contributed to phase diagrams, alloy development, and mechanical behavior of titanium alloys.

The U.S. Air Force at Wright-Patterson Air Force Base, Dayton, Ohio, began extensive in-house and contract programs at this same time. This work was placed at Battelle, ARF, and other laboratories. The Air Force concentrated much of their phase diagram work at ARF, where Max Hansen, an internationally known expert in phase diagrams, was the new manager of metallurgy research.

Hansen was a long-time friend of William Kroll, who was the chief sponsor for Hansen when he obtained his doctorate at the University of Göttingen in 1923. Hansen studied under the famous Professor Gustav Tammann who was a pioneer in phase diagram research. In 1936, Hansen published a massive compilation of phase diagram studies that he collected from worldwide published literature.

Hansen's major field of research was aluminum and the use of aluminum alloys in aircraft construction. He escaped Germany at the end of World War II, avoiding capture by the Russians with British help. He accepted a teaching position at IIT in Chicago in 1947 and transferred in 1949 to the responsibility of manager of metallurgy research at ARF. Under his guidance, IIT began work on titanium phase diagrams and titanium alloy development.

As the phase diagram and alloy development studies funded by the various government agencies were getting underway in 1950 and 1951, a commercial research project on alloy development supported by Rem-Cru was started at Battelle. This program was an investigation of alloys of titanium and aluminum with the addition of a third element. None of the previous work at Battelle included alloy systems of titanium and aluminum, so Battelle was free to perform this study without conflicting with its other government-supported research. Many alloy compositions were examined, and patents were applied for in a broad range of alloys. The most important of these alloys were ones containing titanium with aluminum and vanadium. Later, an alloy in this group became the most important one for aerospace applications.

The widespread research activity in the early 1950s was evidenced by the extent of the contracts with universities and industry. University contracts included Brown (corrosion), Carnegie Institute of Technology (mechanical properties), Case School of Applied Science (mechanical properties), Columbia (phase transformations), MIT (machining), Michigan (machining and heat treatment), Rensselaer Polytechnic Institute (welding), Syracuse (impurities), and Cal Tech (phase studies). Major industrial firms included Allegheny Ludlum (alloy development), Driver-Harris (wire drawing), Fansteel Metallurgical (coatings), A.O. Smith (welding), Sam Tour (chemical analysis), Worcester Pressed Steel (sheet forming), and dozens of firms with smaller activities.

The production of titanium sponge and titanium metal products was slow in getting started. In 1951, the Material Advisory Board (MAB) of the National Research Council, National Academy of Sciences forecast the need for 30,000 tons of metal products; the total shipment was only 75 tons, with 500 tons of sponge produced. This was barely enough metal to supply the research contracts and provide some leftovers to the aircraft industry for evaluation. At the same time, Colonel John Dick of the Air Force was urging aircraft manufacturing companies to use titanium as a replacement for steel. Colonel Benjamin Mesick of the Army Ordnance Corp. placed an order for $1 million worth of metal to help the industry get underway. The problem was not a lack of orders, but the difficulties encountered in melting sponge and fabricating products with this new and unfamiliar material. It was widely believed that titanium could be melted, rolled, and shaped on the same equipment used for stainless steel. To some extent this was true, but the titanium was much more difficult to handle. As orders came in, the steel mills found that they produced more scrap than useful metal.

Even more important for the eventual future of titanium as a structural material was the pricing in 1951 of the ordered product per pound: $15 for sheet, $12 for plate, and $6 for bar. Perhaps the promoters within the armed forces and the other government agencies thought that this was a starting price that would decrease drastically once the increased production levels were reached. Regardless of pricing, the end users within the aerospace industry did not lose their enthusiasm for titanium.

Shipments for 1952 increased to only 250 tons of metal, with production of sponge at 1075 tons. This wonder metal was getting off to a very slow start. The plan was to increase the quantity of sponge production and allow the metal production to follow.

The government provided support for the increase in sponge production by funding projects for construction of new or expanded facilities. The Office of Defense Mobilization (ODM) funded the construction and the General Services Administration (GSA) agreed to buy any surplus product for the national stockpile of strategic materials. Numerous industrial companies responded to this program. The first contracts went to the companies already in the business, TMCA (3600 tons/year) and DuPont (2700 tons/year). Later, contracts were made with the Crane Company (6000 tons/year), Dow Chemical (1800 tons/year), and

Union Carbide Corporation (7500 tons/year). The new production was to meet a goal of 22,000 tons/year forecast to fill the needs of the aircraft and jet engine producers. One additional company, National Distillers and Chemical Corporation, entered sponge production (7500 tons/year) later with its own funds.

Demand for titanium metal products increased in 1953 with encouragement from the Air Force. The aircraft companies were beginning to use titanium in the new fighter and bomber aircraft and Pratt & Whitney was designing it into the latest engines. Producers, however, were on a difficult learning curve and could not keep up with the demand. Total production of titanium metal products was only 1100 tons. This was unacceptable to a defense department trying to maintain support for the troops in Korea and to build what was required for the Cold War with the Soviet Union. The Air Force held a meeting in spring 1953 to discuss the critical situation with all the parties involved. The meeting included representatives from 12 airframe companies, five jet engine companies, and nine government agencies looking for answers from the titanium producers. The result was the realization by all that they could not fulfill every requirement and that priorities had to be established.

Later in the year, a Senate subcommittee under Senator George Malone was formed to investigate the titanium problems. Many witnesses were called before this committee to evaluate the problems and to determine the true value of titanium to the defense effort. Again, the Air Force and the airframe and jet engine representatives were enthusiastic about the need for this new metal. The most vocal witnesses were the high-level executives from Lockheed Martin, North American Aviation, McDonnell Douglas, Boeing, and other aircraft companies. They were predicting plane designs in the near future that would require 40 to 60% of the aircraft weight in titanium. One individual stated that his firm would need 250,000 tons annually in five years. Another put their needs at 800,000 tons by 1960 if the country were again mobilized for war. Added to these predictions were others from the jet engine manufacturers and the Army Ordnance Corps that raised the needs to levels equal to the production of stainless steel.

A voice of caution came from the ODM, which argued that the forecast would require 50% of the total Air Force budget. A lone dissenter from the titanium industry was Edwin Gee of DuPont, who thought that more effort should be spent on finding less expensive processes for making the metal. He reasoned that the cost of titanium as produced would prohibit the general uses being envisioned. The titanium promoters, however, won the day, and new goals for production were established by ODM at 37,500 tons/year of titanium sponge.

A serious technical problem appeared in early 1954 while attention was focused on the future needs and supply problems. The Pratt & Whitney Engine Division and The Douglas Aircraft Company received shipments of metal that was brittle. Sheet metal tore and engine parts cracked under very low stresses. The problem was quickly traced to high hydrogen contents in the metal. Because a minimum hydrogen level had never been specified, and because all the metal at the user's plants and in the distribution pipeline could be contaminated, work on titanium came to a standstill. A massive effort was launched to determine the source of the hydrogen, the safe level for specification, new methods for hydrogen analysis, embrittlement mechanisms, and how to salvage all the metal on hand. Remarkable progress was made by the cooperation of the producers, users, Air Force, Army Ordnance, and the research laboratories and universities involved with titanium.

Vacuum annealing was quickly identified as a process for removing hydrogen from the contaminated metal and the initial panic gradually subsided. The hydrogen problem, however, did not disappear. New tolerance levels were established that required additional vacuum melting and processing for titanium. Melting under vacuum to eliminate gases had been used for titanium in the earliest research at Battelle. Hal Kessler at ARF developed an improved process where he used compacted titanium sponge as an electrode in an arc melting technique. This process was called consumable electrode, vacuum arc melting (VAR). The original idea was invented at Climax Molybdenum Company for use with molybdenum. Kessler adapted it to use titanium sponge. TMCA scaled up the laboratory concept to production size, and this process became standard melting practice after the hydrogen problem (**FIG. 9.5**). Double melting and even

FIG. 9.5 VAR ingot of titanium that will be hot formed into finished product. Source: www.osaka-ti.co.jp.

triple melting was common in titanium for critical applications. The VAR process gradually expanded to other metals, especially nickel-base alloys and iron-base alloys for use in jet engines and other critical aerospace applications.

Attention turned to other difficulties affecting the use and production of titanium. Producers were making progress on improvements in the manufacture and quality of their product, but there was still a requirement for stronger alloys. One aircraft company had complained that they would not design titanium into new planes unless stronger alloys were available. Much of the titanium used previously had been the lower-strength commercially pure titanium (CP titanium). It was easy to process, and early applications substituted it for other metals where heat resistance was more important than high strength. Each producer had developed alloys of higher strength, but these were only moderately strong and were more difficult to fabricate into parts. Fortunately for the industry, an alloy was under development that would solve the problem.

Research Progress

The ARF Metals Division, under the direction of Max Hansen, had been working on a Watertown Arsenal contract for alloy development. Hal Kessler and his group of research engineers were studying various alloy systems, including those containing aluminum and vanadium. One of the most promising alloys contained 6% aluminum and 4% vanadium (Ti-6Al-4V). Sample ingots were supplied to the Arsenal for heat treatment, mechanical property, and ballistic studies. The Air Force initiated a contract at ARF to study the high temperature properties of interest to jet engine applications. Later, ARF supplied 100-pound ingots of Ti-6Al-4V to the jet engine builders for evaluation. The success of this effort soon brought the alloy to the attention of the entire titanium world. Titanium producers immediately began production of the alloy, and before long it was designed into jet engines.

Because the ARF work was done under a Watertown Arsenal contract, the patent belonged to the Arsenal. The Arsenal, however, delayed its patent application because it decided to keep the ballistic information secret. In agreement with Watertown Arsenal, ARF applied for a broad patent on Ti-Al-V alloys, including 6Al-4V. On the basis that a government-sponsored project had resulted in the ARF patent application, one defense contractor ceased paying royalties to Rem-Cru. This precipitated a lawsuit that pitted Rem-Cru against the government and several users and producers. The suit was eventually withdrawn, leaving the invention of Ti-6Al-4V unsettled. To complicate matters further, a patent was granted to Watertown Arsenal for heat treatment of the Ti-6Al-4V alloy. Now the Arsenal, ARF, and Rem-Cru all claimed to have invented the alloy. The final word seemed to be contained in a letter from Charles F. Hickey, Chief, Technology

Management Branch of Watertown Arsenal, to Harold Kessler stating that Kessler was indeed the inventor of the most important alloy in titanium.*

The importance of Ti-6Al-4V cannot be overstated: The titanium industry probably would not have existed without this alloy. An alloy becomes successful when it offers a combination of properties and characteristics that satisfy a wide variety of applications. An alloy does not have to be superior in all properties; a proper balance is more universally valuable. The Ti-6Al-4V alloy was the first titanium alloy to perform this role, and with continued use it became a material that industry felt confident in using. As late as 2000, Ti-6Al-4V represented 75% of all titanium alloys produced. It also was the most important alloy used abroad. Japan, China, England, and Russia produced it. It was used in the construction of the Russian Alfa Class Submarines.

The Titanium Metallurgical Laboratory

Many of the early contracts on titanium science and technology were being completed from 1953 to 1955. Numerous technical papers were being published in a wide variety of journals, each dedicated to a particular specialty, such as welding, corrosion, phase diagrams, and alloy properties. In addition, the metal producers and their customers were developing processes and procedures for their own particular needs. All of this valuable information for the general use of titanium needed to be accumulated, analyzed, and repackaged into documents that could be used by other researchers or by designers and personnel on the shop floor. A $1 million annual contract to perform this effort was awarded by the Defense Department to Battelle Memorial Institute in late 1954. The result was the production, over a two-and-a-half-year span, of 80 technical reports and over 100 technical memoranda to a distribution list of 1800.

The problems with titanium processing, including hydrogen embrittlement, limited its production to 1300 tons in 1954, barely more than in 1953. Sponge production, however, more than doubled, to 5400 tons. Metal shipped in 1955 increased to 1900 tons and sponge to 7400 tons. The main use was in jet engines, and the biggest customer was Pratt & Whitney (**FIG. 9.6**). Gradually, the Ti-6Al-4V alloy was becoming the most important titanium product for the jet engine market.

---

* In response to the April 2015 Advanced Materials & Processes "Metallurgy Lane" article by Charles Simcoe about titanium, Stanley Abkowitz, FASM, compiled a letter with some differing historical evidence. His response is based on a monograph he wrote that includes a detailed discussion of the patent office proceedings that resulted in the Ti-6Al-4V alloy patent. The JOM Monograph, entitled "The Emergence of the Titanium Industry and the Development of the Ti-6Al-4V Alloy—Collections and Recollections," is available from the International Titanium Association (Ref 1).

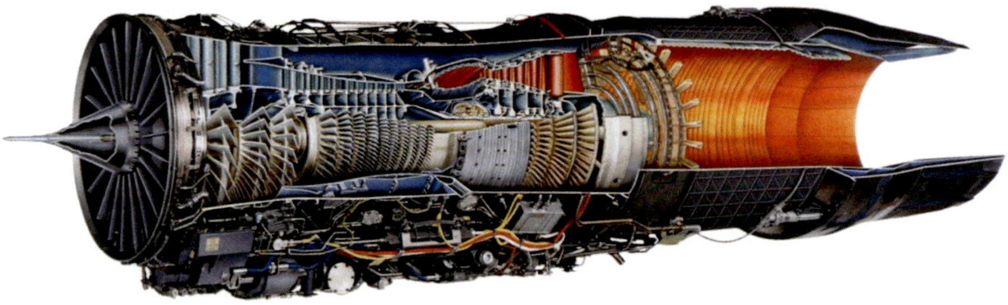

FIG. 9.6 Pratt & Whitney jet engine. Titanium is used for the compression section, where temperatures range from 150 to 540 °C (300 to 1000 °F). Source: United Technologies Corp. 2017.

Sponge production was moving ahead much faster than metal production. In addition, in a case of one hand not knowing what the other was doing, the government had contracted with the Japanese to exchange surplus grains and other foods for titanium sponge. The Japanese sponge that started as a trickle in 1953 reached 600 tons in 1955 and 3600 tons in 1957. The competition forced the per pound price of sponge from $5 in 1953 to $2.25 in 1957.

In 1956, titanium began to show the promise that many promoters had hoped. Metal production reached 5200 tons and sponge production nearly 15,000 tons. The performance yielded profits and attracted new investments and competition into the market. Titanium Metals Corporation of America bought an old steel mill in Toronto, Ohio, and added new specialized equipment for processing titanium. Republic Steel announced a $7 million expansion and bought a 50% interest in the Crane Company sponge plant, now called Cramet Inc. National Distillers began producing sponge (5000 tons/year) without government support. Mallory-Sharon added new plant equipment and other facilities. Two other companies with long-time interest in titanium, Allied Chemical Company and Kennecott Copper Corporation, announced they would jointly build a plant to make sponge and melt ingots. Two companies on the West Coast announced they would begin melting sponge to produce ingots. They were Oregon Metallurgical Corporation in Albany, Oregon, and Harvey Machine Company in Torrance, California.

While the business side of titanium was improving, most of the alloy product was going into jet engines. The airframe manufacturers were still grappling with problems in sheet material. They needed better alloys for forming parts and to improve strength. They also needed more uniform properties. The variation in properties from one producer to another or even within a shipment from the same producer was a cause for concern. The learning curve for sheet products of titanium was more difficult than expected. The Department of Defense (DOD) initiated a substantial project, with a budget of $3.5 million, in 1956 to combat a host of sheet metal problems.

The DOD program on sheet rolling was administered by the National Materials Advisory Board (NMAB). Representatives from all the major airframe manufacturers, jet engine builders, titanium producers, the armed forces, and other government agencies with an interest in titanium assembled at the NMAB headquarters at DuPont Circle in Washington. One of the most comprehensive technical programs ever undertaken in metallurgy was designed by this group to overcome the problems hindering the rapid growth of titanium in airframe construction. Three promising alloys, including Ti-6Al-4V, were selected for study. Each of the three major titanium producers received orders for the three alloys. This material was distributed to the participating airframe companies and to other laboratories for testing and evaluation.

A system of test procedures was designed so that everyone involved tested material from the same source and even from the same sheet. In this manner, discrepancies in the laboratory techniques of each participant were uncovered, as well as the variation in product identified. Periodic meetings were held at NMAB for all representatives to report findings, and these meetings helped guide the program. Over several years, this project produced not only improved alloys and improved material quality, but also a much better appreciation of the reliability of titanium as an aerospace material. An indirect benefit was that close relationships formed between individuals of all parts of the industry through participating in this common goal. This effort could only have been accomplished by using the office of a government agency, in this case the NMAB, to bring all the interests together without regulatory consequences.

The improved business conditions and resulting enthusiasm within the industry continued into 1957. First quarter metal shipments were 2200 tons, which indicated an annual goal of perhaps 9000 to 10,000 tons. Titanium was on its way to becoming the aerospace metal that its promoters had hoped it would be. Then an earthquake shook both the aerospace industry and titanium. The Secretary of Defense, Charles Wilson, formerly of General Motors Corporation, announced a decision to base the defense of the country on missiles rather than on manned aircraft (**FIG. 9.7**). The accompanying reductions and cancellations of contracts rocked the industry. The B-52s were cut from 17 wings to 11, and the contracts for the new Century Series fighters were either cancelled or extended over much longer times. The reduction in aircraft and engine production caused a near-total collapse of the titanium market. This most promising of metals for the postwar era, which had received an estimated $200 million in government support, was in serious danger of extinction by the very DOD that had brought it into being.

FIG. 9.7 Charles Wilson, Secretary of Defense, designed a program to replace manned aircraft with missiles. Source: Historical Office Public Affairs.

The results were swift and drastic. Orders were cancelled, and by the fourth quarter of 1957 shipments slipped to 350 tons. The price of sponge plunged to $2.25 per pound as production increased to 17,000 tons and the United States imported 3500 tons of sponge from Japan in exchange for food. Metal product prices decreased to an average of $10 per pound. These prices continued to decline over the next several years, reaching as low as $1.60 for sponge and $7 for metal.

Many of the early producers were beginning to leave the industry. The Remington Arms Division of DuPont sold its 50% interest in Rem-Cru to its partner, Crucible Steel Company. Crucible remained in metal production for a brief time, then left the field. Cramet Inc., jointly owned by Crane Company and Republic Steel, ceased operation, and Republic stopped production of titanium metal products. Dow Chemical put its sponge plant on standby and shortly closed it. Union Carbide reduced production of sponge to 25% of capacity and later closed. Mallory-Sharon sold a one-third interest to the new sponge producer, National Distillers. Within a few years, both P.R. Mallory and Sharon Steel sold their interests to National Distillers, which was renamed Reactive Metals. In

1964, U.S. Steel Corporation purchased a 50% interest in Reactive Metals. Even DuPont, the first company to enter sponge production, closed its plant in 1962. Witnesses to this situation believed that the DOD program on sheet rolling was a major factor in holding together what remained of the industry.

It was in this low state of the business that a celebration was organized, aimed at honoring the founders on the 10th anniversary of the industry. The individuals listed as founders were those "whose vision, perseverance and effort made possible the birth of the titanium metals industry." For the most part, the founders were involved in the very early years, and many individuals who had made significant contributions as the industry was getting underway in the 1950s were not included.

The changes in the titanium industry had changed the career paths of many of the individuals selected as founders. William Kroll returned to Europe to enjoy his retirement years in Brussels. Max Hansen left ARF to accept a high-level post in the metals industry in Germany. In 1958, Hansen and Kurt Anderko, another German who worked with Hansen at ARF and who had returned to Germany with him, published a monumental second edition of the "Constitution of Binary Alloys."

Two principal military supporters of titanium, Colonel Ben Mesick of Watertown Arsenal and Colonel John Dick of the U.S. Air Force, went on to other careers. Mesick moved to the University of Arizona and Dick joined Allegheny Ludlum. Nat Promisel of the U.S. Navy became head of the NMAB. There was a great fallout of personnel from the many companies who left titanium. A few personnel relocated with the remaining producers, but these companies were downsizing their technical staffs to only 40 to 50% of their former size. Similar personnel losses were experienced in the airframe industry.

By the late 1950s and early 1960s, the major research and development support for titanium by the government had ceased. The founders, as well as the younger researchers, were forced to find other areas of research or move on to other employment. Most of the personnel at Battelle remained there to perform other research. Battelle had a more experienced staff that was better able to find other contract research. Armour Research Foundation, on the other hand, gradually lost all of their titanium research staff, who moved on to universities and industry. NYU eventually left the metallurgical field of engineering, and most of the staff moved to Brooklyn Polytechnic Institute. While many of these individuals continued to participate for a number of years in seminars and other technical meetings to provide their accumulated experience, they were no longer active in titanium research.

Titanium metal shipments would not exceed those of 1957 (5600 tons) until 1962 (6500 tons). During this time, the remaining producers were pressed to find other applications for the metal. The reduced jet engine market was still the major customer, but applications were slowly found in chemical and nuclear plant construction, where corrosion resistance was the primary property. The growing field of missile technology began to consume increasing, although limited, amounts of titanium. The civilian airline industry began using increasing amounts of titanium with the introduction of the new jet-powered aircraft. These planes included the Boeing 707 and the Douglas DC-8. Later models included the Boeing 727 and 737 and the DC-9. In all of these planes, a few percent of the airframe weight was in titanium, as well as the substantial weight in the engines.

One significant development in the 1960s was the building of the SR-71 Blackbird military reconnaissance plane with a speed capability of Mach 3 (**FIG. 9.8**). The SR-71 was an all-titanium aircraft manufactured with Ti-6Al-4V and a new alloy, Ti-13V-11Cr-3Al. Another early 1960s attraction was the interest in a supersonic transport plane. It would have been an all-titanium aircraft that was planned to compete with the European-made Concorde. It was planned to fly at nearly Mach 3 and would have used large amounts of titanium per plane for a planned 200-plane fleet. The program was cancelled by Congress in 1971.

The jet engine builders were the major consumers of titanium, and Pratt & Whitney was the leader. Until the postwar period, Pratt produced only piston

FIG. 9.8 The Lockheed SR-71 Blackbird, built in the 1960s, was the first airplane with all-titanium construction. Source: U.S. Air Force.

engines for aircraft. The transition to gas turbines after World War II was much more difficult for Pratt than for General Electric or Westinghouse, who were long-time manufactures of steam turbines. It was a complete reengineering for Pratt to transfer from reciprocation engines to rotating turbines. They made the transition successfully, whereas Westinghouse Electric Corporation and Curtiss-Wright Corporation, the other major piston-engine manufacturers, dropped out of the field. Pratt & Whitney were the first to embrace titanium in their military jets in the early 1950s. Later, when commercial jet engines became popular in the 1960s and beyond, they continued to expand the use of titanium.

Eli Bradley was the chief materials engineer at Pratt during this crucial period of the development of commercial jet engines. For 20 years, he was a leading expert in titanium applications in the jet engine industry. Bradley was raised in Connecticut and graduated from Yale in metallurgical engineering in 1939. He joined Pratt & Whitney out of college and worked there his entire career. Bradley received the ASM Engineering Materials Achievement Award in 1975 and was given the highest honor in metallurgy, the ASM Gold Medal, in 2002, the same award that had been given to William Kroll in 1965.

Introduction of the wide-bodied air transport planes in the late 1960s improved the markets for titanium. Each Boeing 747, for example, consumed 15 tons of titanium products for the airframe and approximately 25 tons for the engines. After a decline in shipments in the early 1970s, business increased with the construction of newer fighter planes and with improving markets in the chemical and nuclear industries. The new Grumman F-14 Tomcat used 35% of the airframe weight in titanium and the McDonnell Douglas F-15 Eagle used 24%. Shipments improved by 1980, to an all-time high of 27,000 tons. A number of factors coincided to push the use of titanium to nearly the 30,000 tons forecast in the early 1950s. Civilian transport aircraft construction and chemical and nuclear uses reached peak levels, and even export business had developed to serve the growing military and civilian aircraft production in Europe. This peak usage would not be repeated for many years due to the rapid decline in all of these markets in 1982–83. Not a single nuclear plant was built in this country after the Three Mile Island accident in 1979. Chemical plant construction went into a long decline, and aircraft sales leveled off during the recession in the early 1980s. The one bright spot was the B-1 Lancer bomber built by North American Rockwell (**FIG. 9.9**). Each of the 100 planes built required 100 tons of titanium.

The most recent commercial aircraft designs use substantial amounts of titanium. The Boeing 777 uses 13,000 pounds of Ti-10V-2Fe-3Al in the landing gear (**FIG. 9.10**). Ti-10V-2Fe-3Al is a beta alloy that is heat treated to 160,000 to 170,000 psi. Other alloys used in a variety of applications for this plane include commercially pure titanium, Ti-3Al-2½Sn, Ti-6Al-4V, Ti-6Al-2Sn-4Zr-2Mo-2Si, and a beta alloy Ti-3Al-8V-6Cr-4Mo-4Zr. This latter alloy is heat treated to 200,000

FIG. 9.9 The B-1 Lancer bomber was built to withstand speeds of 2000 mph. The heat generated required titanium alloys, mainly Ti-6Al-4V.

psi. In addition to use in the airframe, the Boeing 777 includes approximately 25,000 pounds of titanium alloy in the engines.

The Lockheed Martin F-22 Raptor uses approximately 42% (9000 pounds) of titanium alloys in the airframe, although the largest amount is still Ti-6Al-4V. The

FIG. 9.10 The Boeing 777 is the first commercial plane to use a titanium alloy (Ti-10V-2Fe-3Al) for landing gear. Source: Wikimedia Commons/Altair78.

Pratt & Whitney engines for the F119-PW-100 contain Ti-6Al-4V and the newer alloy Ti-6Al-2Sn-4Zr-2Mo-0.2Si.

A few of the original titanium founders attended the 1960 celebration of the 10th anniversary, and they formed a small committee in the late 1980s to discuss writing about their experiences in the development of the industry. They included Nate Promisel of the Navy and the NMAB, William Harris of the NMAB and Battelle, Robert Jaffee of Battelle, Hal Kessler of ARF, TiMet, RMI, and other producers, Robert Nycum of TiMet, Ward Minkler of TiMet, and Andrew Eshman of RMI. This group met over several years while attending other major metallurgical meetings. Their goal was to combine their experiences, along with other information, into a book on the complete history of the development of titanium. It was an ambitious undertaking. A brief outline was prepared, and one or two individuals wrote their sections, but others held back until arrangements were made for a firm commitment for financial support by the titanium industry. This help never materialized, so most of the history was never written.

Titanium is familiar to many Americans in consumer items. It is available in jewelry, wristwatches, golf clubs, eyeglass frames, bicycle and auto engine parts, and handguns. A growing application is in the biomedical field, where it is used in prosthetic knee and hip joints (**FIG. 9.11**).

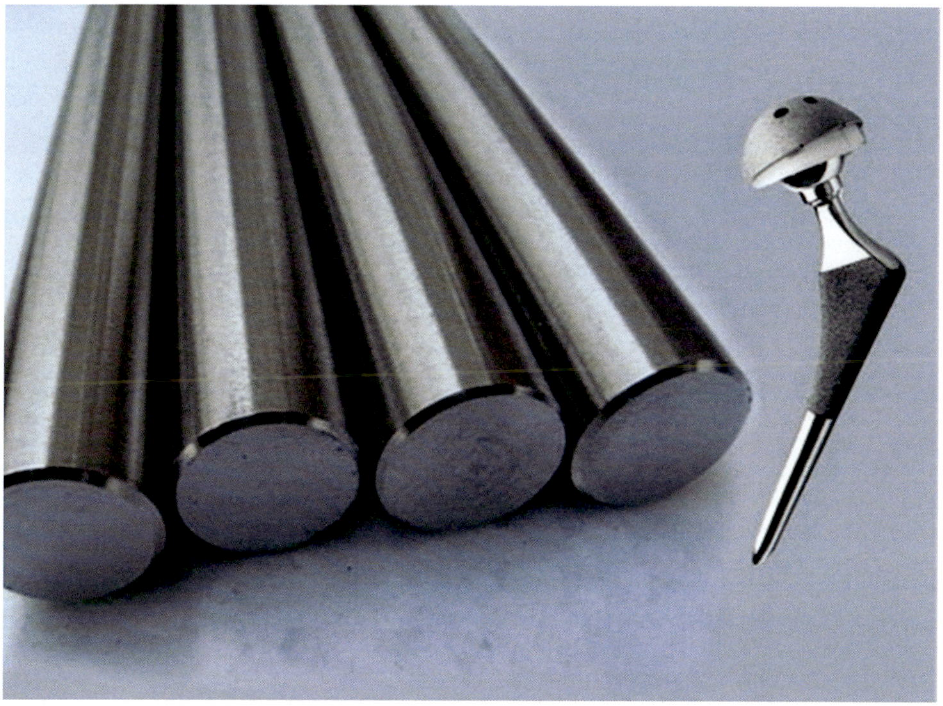

FIG. 9.11 Finished bars of titanium that are forged into joint replacement parts. Source: www.surgicalimplanttitanium.blogspot.com.

These consumer uses were not anticipated in the late 1940s and 1950s, when titanium was seen as a future structural material for defense applications. However, it required 50 years for production to reach 30,000 tons per year, a goal that had been forecast for 1960 by the early studies.

The original enthusiasts, individuals, companies, government agencies, universities, and research laboratories believed in a great future for a strong, lightweight, corrosion-resistant metal. This future came to pass after a decade of difficult research and development and an investment of several hundred million dollars by the government and industry. The cost of producing titanium, however, limits its major applications to jet engine and airframe construction and some chemical corrosion uses. For these uses, titanium alloys are indispensable. The modern fan bypass jet engine would not be possible without titanium.

REFERENCES

1. S. Abkowitz, *The Emergence of the Titanium Industry and the Development of the Ti-6Al-4V Alloy*, The Minerals, Metals and Materials Society, JOM Monograph Series, Vol 1, 1999

Selected References

E. Bradley, The Role of Titanium in the History of the Jet Engine, *Journal of the Japan Titanium Society*, 1982

M.J. Donachie, *Titanium: A Technical Guide*, American Society for Metals, 2000

F.H. Froes, D. Eylon, H.B. Bomberger, *Titanium Technology: Present State and Future Trends*, Titanium Development Association, Dayton, OH, 1985

K.L. Housley, *Black Sand: The History of Titanium*, Metal Management Aerospace, Inc., Hartford, CT, 2007

W.J. Kroll, How Commercial Titanium and Zirconium Were Born, *J. Franklin Instit.*, Vol 260 (No. 3), Sept 1955, p 169–192

J.W. Marden, William Kroll–Medalist, *J. Electrochem. Soc.*, Vol 106 (No. 1), 1959, p 7C–8C

E.R. Rowley, and P. Cloyes, *A History of Titanium Sponge Manufacture*, Manufacturing Chemist Association, Washington DC, 1960

R. Watercott, *Titanium: Part Product, Part Cause*, Titanium Founders Meeting, Washington, D.C., 1960

Chapter Ten

Pioneers in Metals Research

Henry Marion Howe

Henry Marion Howe was an industrialist, scientist, teacher, member of the National Academy of Sciences, and always a writer (**FIG. 10.1**). He wrote his first textbook in 1893 on the subject of steel as it was known at the time. He performed research all of his life, and even in retirement, he worked in a laboratory at home.

Howe is best known for teaching metallography to many students who went on to develop modern physical metallurgy. He started teaching at the

FIG. 10.1 Henry Marion Howe.

Massachusetts Institute of Technology (MIT) after a brief career in industrial metallurgy. He later accepted a professorship at Columbia University and was the first professor of metallurgy in the School of Mining Engineering in 1897.

Howe was born in 1848 to parents who were well known in Boston's intellectual circles. His father, Samuel Howe, was a physician who established the Perkins School for the Blind (**FIG. 10.2**). His mother, Julia Ward Howe, was a poet and ardent supporter of civic causes (**FIG. 10.3**). She was famous as the author of the poem that became the lyrics of the Civil War song, *The Battle Hymn of the Republic*. She was also in demand as a lecturer in the women's suffrage movement. Members of the Howe family accomplished things with passion, and Henry Marion Howe brought this passion to the field of metallurgy.

Howe's early education was with tutors from Europe, where his father had many connections after serving in the Greek army during the war with the Turks. From these tutors, Howe learned French and German. He attended high school at the famous Latin School in Boston and enrolled at Harvard in 1865, where he received a liberal arts education. Much to his family's dismay, he then studied mining engineering at the recently founded MIT, graduating in 1872. As part of his student training, he worked for a year at the Albany and Rensselaer Iron & Steel Company in Troy, New York, on the Bessemer steel process. Here he came

FIG. 10.2 Samuel Howe, Henry's father.

FIG. 10.3 Julia Ward Howe, Henry's mother.

in contact with Alexander Holley, who was building a steel industry with this new process.

After graduation from MIT, Howe accepted a job as superintendent of the new Bessemer plant at Joliet Iron and Steel Company in Illinois. He stayed only one year and then accepted a job in Pittsburgh at the Blair Iron and Steel Company. After one year in this position, he returned to Boston to spend time writing. In 1877, he entered an entirely new field when he went to Chile to oversee the copper mining operations of a wealthy Boston family. Returning to the United States in 1879, he supervised the construction of copper smelters in Canada and New Jersey. He then became manager of a copper smelting company in Pima, Arizona. After one year, he again returned to Boston, ending his career as an industrial metallurgist. Howe set up a consulting practice and began teaching at MIT.

Over the next several years, he worked on his first major book, *The Metallurgy of Steel*, which was published in 1890 (**FIG. 10.4**, Ref 1). This book established his reputation as a leading metals engineer throughout the industrial world. Howe surveyed all of the published literature on steel from the United States, England, France, Germany, and Russia to develop the book. The bulk of the text covered the processes of steelmaking and included very little on what is now called physical metallurgy. He also covered the known information on alloys, including chromium, manganese, nickel, and tungsten. Much of the alloy material he wrote about included information on mechanical properties. In most instances, he found very little difference between these alloys and plain carbon steel.

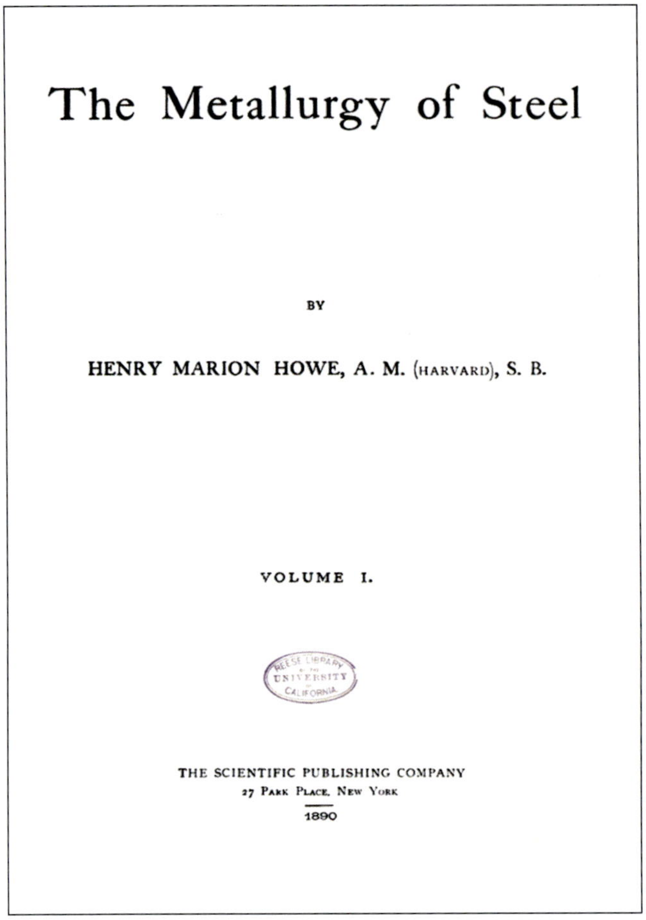

The Metallurgy of Steel

BY

HENRY MARION HOWE, A. M. (HARVARD), S. B.

VOLUME I.

THE SCIENTIFIC PUBLISHING COMPANY
27 PARK PLACE, NEW YORK
1890

FIG. 10.4 First page of *The Metallurgy of Steel*, published in 1890.

He offered extensive information about chromium in steel. He concluded through chemical analysis that most "chromium steels" actually contained little or no chromium. He blamed this fact on poor workmanship. He cited the Eads Bridge across the Mississippi as an example. Chemists found little or no chromium in the steel even though the bridge was supposed to be made of crucible steel that contained chromium. (Recent studies of the steel in Eads Bridge found 0.60% chromium and 0.60% carbon.) The steel for Eads Bridge was made by the crucible process in 100-pound batches, so the chromium content varied wildly. He finally concluded that there was evidence that chromium could, in amounts of 2%, increase the depth of hardening in large parts. He discovered the secret of alloy steels: They have the ability to increase what we now call hardenability. However, neither Howe nor anyone else followed up on this important concept, which would not be resolved until the 1930s.

Howe joined the Columbia School of Mines in 1897 as the first professor of metallurgy. He taught the new field of metallography and wrote *Metallurgical Laboratory Notes* (1902) and two more major books, *Iron, Steel and other Alloys* (1903) and *The Metallography of Steel and Cast Iron* (1916). He retired in 1913 at age 65 and moved into his second home, called Green Peace, in Bedford Hills, New York (**FIG. 10.5**). His home was equipped with a full laboratory, where he continued his research.

Throughout his career, Howe was active in technical organizations. He joined the newly founded American Institute of Mining Engineers in 1871, along with Alexander Holley, while a student in Troy, New York. He was elected president of that society in 1893. He was very active in organizations working on materials standards and joined the International Society for Testing Materials. Later, he was a founder of the American Society for Testing Materials (ASTM) and served as president from 1909–1911. Howe also joined the new Steel Treaters Club (later ASM International) when it was formed in 1913.

With the entry of the United States into World War I, Howe came out of retirement and volunteered his services to the federal government. He served as a consultant to the Bureau of Standards and the Bureau of Mines and was appointed to the National Research Council. After the war, he retired to his home in Bedford Hills where he died in May 1922 at age 74.

FIG. 10.5 Henry Marion Howe's "Green Peace" home and laboratory in Bedford Hills, New York. Source: University of Pennsylvania.

Howe's reputation was built on his trio of books on steel. His legacy, however, was based on his metallography courses. They attracted students from across the country including Edgar Bain and Earl Smith from Ohio State, Samuel Hoyt from Minnesota, Joseph Emmons from Cleveland, and many others who would go on to teach the subject or do major steel research. Among his many honors was the Henry Marion Howe Medal, established in 1923 by ASM for his lifetime contributions to steel metallurgy (**FIG. 10.6**). The Howe Memorial Lecture was also established in 1923 by the Association for Iron & Steel Technology. His awards include the Bessemer Medal of the Iron and Steel Institute of England, The Legion of Honor from France, Order of the Knights of St. Stanislaus from Russia, The Gold Medal from the Societe d'Encouragement pour L'Industrie Nationale of France, the Elliot Cresson Medal from the Franklin Institute, and the John Fritz Medal from the American Association of Engineering Societies. He was selected to the National Academy of Sciences in 1917.

Albert Sauveur

Albert Sauveur was the first metallurgist in America to study the microstructure of steel for a plant laboratory. He worked for five years, from 1891 to 1896, on the metallography of railroad rails for the Illinois Steel Company in Chicago.

FIG. 10.6 ASM's Henry Marion Howe Medal, established in 1923.

Sauveur was born in 1863 in Louvain, Belgium, but was of French ancestry. He studied at Athenee Royal in Brussels from 1881 to 1886. He immigrated to the United States to attend the Massachusetts Institute of Technology (MIT), where he received a Bachelor of Science degree in mining and metallurgy in 1889 (**FIG. 10.7**). His first position was as a chemist at the Steelton plant of the Pennsylvania Steel Company. In 1891, he joined the Illinois Steel Company as a metallographer. There he spent five years studying the metallography of railroad rails.

He published his first research paper, "The Microstructure of Steel," in the *Transactions of the American Institute of Mining Engineers* in 1893. He then published work in the *Engineering and Mining Journal* with Henry Marion Howe, entitled "The Relations between Temperature and Grain of Steel." He and Howe published several more papers over the next few years. Their collaboration stemmed from their acquaintance at MIT, where Howe was an instructor.

FIG. 10.7 Albert Sauveur served most of his career as a professor of metallurgy at Harvard University. Source: National Academy of Sciences.

With his European background, he maintained communication with the major metallographers in England, France, and Germany. These were the metals scientists who were performing the first studies of the internal structure of metals. In 1896, he returned to Boston, where he set up a private research laboratory and then joined Harvard University in 1899 as an instructor of metallurgy. He served the rest of his career there as associate professor and professor, and he finally won the Harvard Gordon McKay Professor Award for excellence in research and teaching in 1924.

He started a quarterly journal in 1897, *Metallographist*. It had a short life, but it was instrumental in creating interest in metallography. He wrote many articles for the journal, especially biographies of early researchers. As a professor, he started a correspondence course in metallography that enrolled 1500 students. He wrote his well-known book, *Metallography and Heat Treatment of Steel*, which went through four editions by 1937, with 20,000 copies sold. During World War I, he consulted on alloys for aircraft engines with the U.S. government.

Sauveur received many honorary degrees for his research and teaching, from Case School of Applied Science in Cleveland; University of Grenoble in France, Lehigh University in Bethlehem, Pennsylvania; and Harvard University in Boston. His highest honor was the Franklin Medal from the Franklin Institute in Philadelphia (**FIG. 10.8**). Other recipients include Thomas Edison, Albert Einstein, Stephen Hawking, and Bill Gates. He received the Bessemer Medal from England and was the first recipient of the Albert Sauveur Medal in 1934 from the American Society for Metals (ASM). He was also elected to the National Academy of Sciences.

FIG. 10.8 Albert Sauveur received the Franklin Institute Medal posthumously in 1940.

Isaiah (Zay) Jeffries

Isaiah (Zay) Jeffries was the first graduate in a second generation of students to make a life career in metallurgy. He studied mining engineering at the South Dakota School of Mines. After a brief time working in the mining industry, Jeffries spent his career working in metallurgy. During the 1920s, he became the most widely known metallurgist in America, for his work on tungsten for lamp bulbs, his research into grain size and its influence on mechanical properties, his invention of many alloys of aluminum, and his work on the behavior of high-speed tool steels during secondary hardening. He also trained many young metallurgists who worked with him at his laboratories in Cleveland.

Jeffries was born in the Dakota Territory in 1888, a year before it became the state of South Dakota. His parents, Johnson and Florence Jeffries, had moved west from Illinois for farming and ranching opportunities. Zay grew up in Fort Pierre, a small community on the Missouri River. As a teenager, he helped on the farm, mainly capturing wild horses and breaking them for riding. He attended high school in Pierre, a larger community across the Missouri River, where he discovered an interest in geology, which he hoped to pursue in college (**FIG. 10.9**).

Jeffries enrolled in 1906 at the new South Dakota School of Mines in Rapid City. However, he found that the education and opportunities in geology were limited, and he soon transferred to mining engineering. Jeffries received metallurgical and metallographic training from the university's president, Professor Charles A. Fulton, who had studied under Henry Marion Howe at Columbia

FIG. 10.9 Zay Jeffries was the first of his generation of pioneers to teach, consult, and perform research for industry.

University. Jeffries graduated in 1910 in a class of 10 out of a total enrollment of 44 students. After working in the mining industry for a year, he received an invitation from Fulton, who was then a professor of metallurgy at the Case School of Applied Science in Cleveland, to join him as an instructor of metallurgy and metallography. He immediately accepted the offer and moved to Cleveland in 1911.

After four months at Case, Zay returned to Rapid City and married Frances Schrader, his girlfriend of college days. They returned to Case in time for the new semester in the winter of 1912. For the next two years, he devoted all his time to teaching and learning more about metallurgy and metallography. By 1914, Jeffries felt prepared for consulting assignments.

His first major assignment was with General Electric's Lamp Division, which was transitioning to tungsten filaments for light bulbs. He successfully solved several problems involving microstructure, especially with regard to grain size. Jeffries' most influential contribution was his research on grains and grain size and their impact on the mechanical properties of metals. Of special interest at the time was what held grains together—that is, what filled the space between grains where different orientations meet.

A theory advanced at the time was that this space was filled with *amorphous metal,* metal without crystal structure. This amorphous metal acted as a cement that held grains together during plastic strain, such as hot and cold working. Elongation of grains during cold working was believed to produce amorphous material—until a few years later, when x-ray diffraction uncovered that no matter how far grains are distorted, they still retain their crystalline structure. As late as 1924, Jeffries could find no solution to grain boundaries other than the amorphous cement theory.

A second consulting assignment that Jeffries undertook was with the Aluminum Casting Company in Cleveland. The company was having problems with casting defects in parts being exported to countries allied against Germany in World War I. Even though the company was partially owned by Alcoa, they did not have the knowledge to help their customers with technical problems. Jeffries became the expert in aluminum metallurgy.

In 1917, he ended his teaching career and enrolled in a graduate degree program at Harvard University under the well-respected Albert Sauveur. He spent a year in residency at Cambridge with his family, then returned to Cleveland to finish his degree by publishing his work *Grain Structure in Metals.*

In 1920, Alcoa acquired the Aluminum Casting Company laboratory for nonpayment of aluminum purchases. Jeffries was now a consultant to Alcoa. He was joined by Robert Archer, a recent chemical engineering graduate from the University of Michigan. Jeffries and Archer worked on improved alloys for forging and castings during the 1920s, sharing patents and papers on all their work.

FIG. 10.10 Cross-sectional view showing automotive engine pistons made of a high silicon alloy developed by Jeffries and Archer. Source: Wikimedia Commons/Mj-bird.

During this time, they invented casting alloys using 5 and 7% silicon (355 and 356), an aluminum-copper alloy (195), an aluminum-magnesium alloy (220), and an automotive piston alloy (132) containing 12% silicon (**FIG. 10.10**).

They developed a new forging alloy by adding 0.75% silicon to Duralumin (14s) and a copper-free alloy containing magnesium and silicon (25s). Alloy 25s was used to make aircraft propellers during World War II (**FIG. 10.11**). They shared in 10 patents and 13 published papers during this decade and coauthored the book *The Science of Metals* in 1924.

This assignment, along with his work at General Electric, lasted throughout the 1920s. Jeffries was joined by many new metals researchers, who trained under him, then made a reputation of their own in later years. One of these associates was Edgar Bain, who worked with Jeffries on problems with high-speed tool steels that failed during the drawing of tungsten wire. They published a research paper on slip interference by alloy carbides for secondary hardening of tool steels. This theory was accepted until the discovery of dislocations was found to be the mechanism for plastic deformation.

FIG. 10.11 Jeffries and Archer developed a forging alloy for propellers used on World War II aircraft, such as the Republic P-47N Thunderbolt shown here. Source: U.S. Air Force.

Jeffries remained at Alcoa until 1936, when he left to assume full-time duties at General Electric. His work at Alcoa became a part-time assignment while he continued to serve as a consultant at General Electric. This dual work arrangement provided him with an income well above that of most engineers of the time.

His new full-time career at General Electric included primary responsibility for developing a commercial business in tungsten carbide tools. Tungsten carbide later became a tool material that was ideal for severe machining applications, and Jeffries played a large role in promoting the business into a major industry. After World War II, Jefferies was appointed vice president of General Electric's new chemical division.

Jeffries' career awards for his advancement of the science and technology of metals included the James Douglas Gold Medal by AIME in 1927, the Albert Sauveur Achievement Award by AIME in 1935, ASM Honorary Membership in 1939, the first Gold Medal Award by ASM in 1943, the Clamer Silver Medal by the Franklin Institute in 1945, the Powder Metallurgy Medal by Stevens Institute of Technology in 1945, and the John Fritz Medal in 1946 (**FIG. 10.12**). He was

FIG. 10.12 Zay Jeffries received the John Fritz Medal for his career in metallurgy.

elected to the National Academy of Sciences, Engineering Division, in 1939, and served as a consultant to Arthur Compton in World War II on the Manhattan Project. For his wartime work, Jeffries was awarded the Medal for Merit, the highest civilian honor by the U.S. government.

During his active career, he played leading roles in several technical organizations, including the American Society for Metals, the American Institute of Metallurgical Engineers, the American Society of Mechanical Engineers, and others. Among his peers, Jeffries was considered the elder statesman of American metallurgy. After retirement, he served as chairman of the ASM World Congress in 1951 and again in 1956. Jeffries died of cancer in 1965 at the age of 76.

Paul Dyer Merica

Paul Dyer Merica and his associates at the National Bureau of Standards in Washington discovered the principles that governed the age hardening of an aluminum alloy called Duralumin. This aluminum-copper-magnesium alloy was invented by a German, Alfred Wilm, in the early 1900s and was in production during World War I in Germany, France, and England. Germany built nearly 100 giant Zeppelins using Duralumin for the framework. France and England used it for limited aircraft parts. The United States was behind in both military aircraft and Duralumin development, so it did not use the alloy in these early days.

The metallurgical world believed Duralumin was a unique alloy that could not be duplicated in any other alloy system. The manner in which it aged to high strength was a complete mystery. In this atmosphere, the National Bureau of Standards initiated a research program to study the alloy's behavior. Paul Dyer

Merica was selected to head a small group of researchers to study the behavior of Duralumin during age hardening.

Results showed that aging in Duralumin was not unique, and it became the basis for a new strengthening mechanism in metal alloys that could be applied to many alloy systems. The principle of age hardening is considered the most important metallurgical discovery in the first half of the 20th century.

Paul Dyer Merica was born in 1889 to Charles and Alice Merica. He was raised and educated in Warsaw, Indiana, and attended DePauw University in Greencastle, Indiana, where his father had preceded him some 20 years earlier (**FIG. 10.13**). After three years, Merica transferred to the University of Wisconsin in Milwaukee, where he received a degree in physics. He taught physics at Wisconsin for one year and then left the United States for two years to teach at Zhejiang University in China.

In 1911, he enrolled for graduate studies at the Royal Institute of Technology in Charlottenburg, Germany. He was a classmate of Samuel Hoyt, an American who had entered metallurgy at Columbia University, in a course taught by William Campbell. Hoyt, in his book *Men of Metals*, described Merica's appearance

FIG. 10.13 Paul Dyer Merica led the team that discovered the principles governing precipitation hardening.

when he arrived: "One day a young man presented himself at the laboratory and announced he might be joining us. His attitude seemed to say that he might just possibly if he liked what he saw. He wore a straw hat obviously too small worn on the side of his head," (Ref 2). Merica and Hoyt had a number of classes together, including one course taught by Adolf Martens, for whom martensite was named.

Merica worked for a brief time at the University of Illinois when he returned with his Ph.D. in physics and metallurgy. He then joined the National Bureau of Standards, where he worked for an older physicist and metallurgist, George Burgess, a Massachusetts Institute of Technology graduate who also studied under the famous Henry Louis Le Chatelier at the Sorbonne in Paris. Merica led a group of metallurgists in studying the mechanism responsible for age hardening in Duralumin. His fellow workers included Howard Scott, R.G. Waltenberg, and J.B. Freeman.

The study was designed to examine the role of Cu and Mg independently, and then to combine them in various amounts to determine how they interacted to cause age hardening. Sample alloys were made at the New Kensington, Pennsylvania, plant of the Aluminum Company of America. They were rolled, annealed, and rerolled until they were reduced to 0.813 mm (0.032 in.) thickness in the cold-rolled condition. At the Bureau of Standards, they were tested in the cold-rolled condition and also tested after annealing to soften them to a baseline measurement. Other samples were solution-treated at high temperatures up to 520 °C (970 °F) and water-quenched to room temperature. They were then aged for 11 days at room temperature or for three days at 100 °C (200 °F). All samples were tested for hardness, ultimate tensile strength, and elongation.

The alloys with Mg up to 3.5% and no Cu showed no age hardening, even though there was precipitation of MgAl compound with long aging times at the higher aging temperatures. The copper-containing alloys with up to 3.2% Cu exhibited an increase in ultimate tensile strength, from 150 MPa (22,000 psi) as quenched to 205/240 MPa (30,000/35,000 psi) after aging, both at room temperature for 11 days and at 100 °C after three days. After longer aging times at higher temperatures, a precipitate of $CuAl_2$ was visible in the microscope, and the mechanical properties decreased. Alloys with copper and magnesium in combination aged to ultimate strengths of 275 to 415 MPa (40,000 to 60,000 psi), depending on the various amounts of copper and magnesium present in the alloy. The higher strengths were after aging at 100 °C and with 3 to 4% Cu and 1 to 1.25% Mg.

To supplement this research, the team investigated the solubility of copper in aluminum. They found that 4% was soluble at 500 °C (930 °F) but decreased to 1% at 300 °C (570 °F) and continued to decrease at lower temperatures. This decrease in solubility was the basis for the age hardening in Duralumin. Rapid quenching produced a highly supersaturated solid solution that rejected copper at a temperature where diffusion took place. The discovery of Duralumin was fortuitous because diffusion occurred at room temperature.

The authors stated, "If the temperature of the alloy is raised to 100 °C, or even to ordinary room temperature, according to the theory which the authors propose, the mobility of the molecules becomes sufficiently great that precipitation of the $CuAl_2$ takes place in the form of very fine particles of colloidal dispersion."

The authors concluded, "Upon aging a quenched sample at 200 °C (390 °F), the hardness first increases to a maximum and afterwards decreases. During that aging there has been first a formation of fine nuclei of $CuAl_2$ followed by coalescence of these particles into ones of larger size. There is, therefore, a certain average size of particle of $CuAl_2$ for which the hardness is a maximum; atomic dispersion of the solute is not the dispersion that produces the maximum hardness but some intermediate one between it and that at which the particles become visible by ordinary means. To this dispersion is due the hardening of duralumin."

The results of this study by Paul Dyer Merica et al. were quoted by Zay Jeffries as "the shot heard around the world." If Duralumin could be hardened by precipitation of a compound, then this new hardening mechanism could be applied to many other alloy systems. The resulting developments opened the door to a whole new world of strong alloys.

Merica received many awards for his published papers on precipitation hardening in Duralumin. They included the John Fritz Medal, the Platinum Medal from the British Institute of Metals, ASM's Gold Medal, a medal from the Franklin Institute, and the AIME James Douglas Gold Medal. Merica became a member of the National Academy of Sciences in 1942.

He joined the International Nickel Company in 1919 in New York after his five years at the National Bureau of Standards. During the next 38 years, he advanced from research metallurgist to president of the parent company, the largest nickel company in the Western world. He died of a heart attack in 1957 at the age of 68.

Edgar C. Bain

Another researcher from the generation of the graduating classes of 1910 to 1915 was Edgar C. Bain (**FIG. 10.14**). He became well known for his work on high-speed steel, the effect of chromium in steel, and his pioneering work using x-ray diffraction to define the crystal structure of solid solutions. His most important research was the study of the transformation of austenite to lower-temperature

FIG. 10.14 Edgar C. Bain's work on isothermal transformation showed a new phase, which was named bainite in his honor.

phases during isothermal transformation. This unique research made him world famous and led to a new phase, bainite, named in his honor.

After high school in Marion, Ohio, Bain enrolled in 1908 at The Ohio State University (OSU) in chemical engineering. His initial interest in metallography began in a class where he saw photomicrographs of ferrite, pearlite, and martensite. Bain followed this field of science throughout his career. His first employment after graduation was with the National Bureau of Standards in Washington. During his short time at the Bureau, he became acquainted with Howard Scott and Paul Merica of precipitation hardening fame. After several years, he returned to OSU to work on an advanced degree, and he took the only course offered in metallography and metallurgy. Before he earned his master's degree, the head of the department recommended him as an instructor at the University of Wisconsin teaching metallography and pyrometry. Due to his limited knowledge in both subjects, it was recommended that he enroll in a summer course in these subjects. He selected the laboratory course at Columbia University taught by William Campbell and Henry Marion Howe.

He taught at Wisconsin for one year, then accepted a research position at the B.F. Goodrich Company in Akron, Ohio. When the United States entered World War I, Bain joined the army, where he worked in chemical warfare research. After his discharge, he joined the General Electric Lamp Division, where he worked under Zay Jeffries, the country's leading expert in tungsten, aluminum, and the effect of grain size on the behavior of metals.

His first assignment was to investigate the problem of failed dies of high-speed steel during the wire drawing of tungsten. He studied the mechanism of secondary hardening, which was still a mystery after 20 years of use. Bain and Jeffries published their results in *Iron Age* magazine (1923) in a famous paper entitled, "Red Hardness of High-Speed Steel." They combined the principles of precipitation hardening discovered by Merica with a mechanism of slip. The secondary hardening they proposed was that the formation of tungsten carbide at 595 °C (1100 °F) keyed the slip to increase the hardness.

Other research conducted by Bain included the study of the crystal structure of metal solid solutions using the new tool of x-ray diffraction. Previous theory, taught by Howe at Columbia and Sauveur at Harvard, stated that there were both solvent and solute patches of crystal structure. Bain's results for combining copper and zinc (brass) showed for the first time that the solute atoms of a different crystal structure simply replaced solvent atoms at random without changing the crystal structure.

After four years working with Jeffries at GE, Bain joined Atlas Steel in Dunkirk, New York, in 1924, where he worked on high-speed and other alloy tool steels with Marcus Grossmann (**FIG. 10.15**). Over a period of one and a half years, they

FIG. 10.15 Marcus Grossmann, left, and Edgar Bain of Atlas Steel Corp., Dunkirk, New York, circa 1923.

published four papers on their research. This was an especially productive time for Bain, as he was pursuing a career studying transformation during the heat treatment of steel and the effect of alloy additions. Alloy steels were becoming ever more important with the expanded production of automobiles and farm and construction machinery, but the understanding of heat treatment and alloying elements had made little progress.

In July 1924, Bain left his good friend Marcus Grossmann and Atlas Steel to join the Union Carbide and Carbon Corporation on Long Island, New York, a producer of ferroalloys, including ferrochromium. He worked on the iron-chromium alloy system, which included studies by x-ray diffraction, his favorite tool.

During the summer of 1927, he took a leave from his work and, with his friend Grossman and their wives, made a trip to Europe. He and Grossman visited steel plants, laboratories, and universities, and the wives toured museums, art galleries, and cathedrals. In his biography, *Pioneering in Steel Research*, he described their visits to many facilities (Ref 3). They met with some of the early research workers who had done the first studies on tool steels, alloy steels, and the transformation of austenite. Bain was disappointed that he missed meeting Albert Portevin in Paris, who was on vacation at the time. He greatly admired the research published in 1919 by Portevin and Gardenar on the transformation of iron and referred to this work throughout his active career. He finally did meet Portevin on a trip he took to Paris in 1935.

Bain joined the newly formed Research Laboratory of U.S. Steel Corp. in Kearny, New Jersey, in 1928. His first assignment was to design and equip a laboratory in a vacant building. Here he undertook research on the transformation of austenite to pearlite in a 0.80% carbon steel. This steel had the carbon content that formed only pearlite and exhibited the eutectoid composition. He hired a young metallurgist, E.S. Davenport, to help with his research.

The experiment was unique in that they studied the formation of pearlite over time at a constant temperature. This was the first time anyone had studied a metal reaction as a function of time. It had been assumed that only temperature was important in transformation. Their experimental approach was to heat very thin samples to the austenitic phase and quench them in a second bath heated to the transformation temperature. At various time intervals, a sample was removed and quenched in a room temperature water bath to form martensite in the untransformed austenite. The microstructure was then examined to measure the amount of pearlite that had formed at the constant higher temperature.

Plotting temperature of transformation on the vertical axis against the time on a logarithmic scale for transformation on the horizontal axis provided the rate of transformation. The resulting curves started slowly, progressed rapidly,

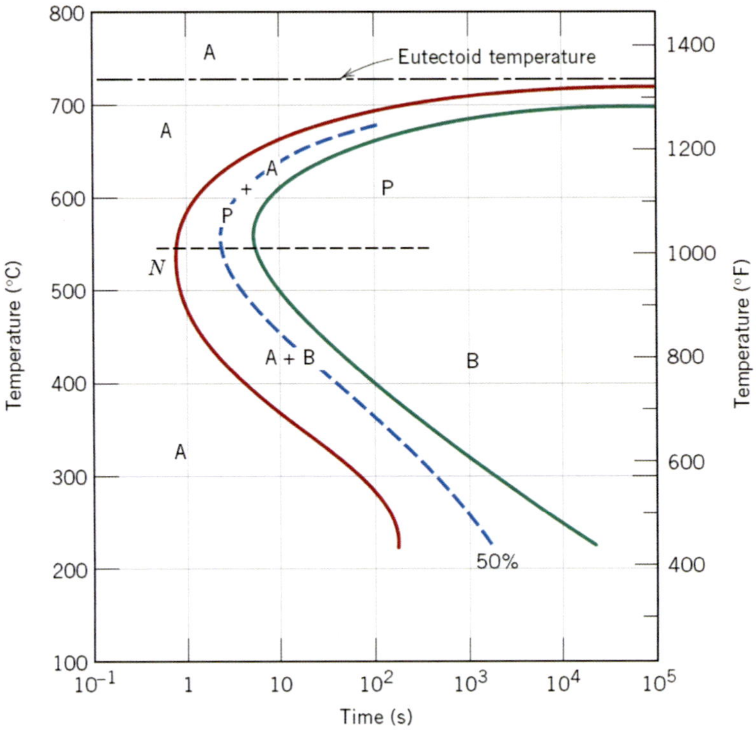

FIG. 10.16 Isothermal transformation diagram for an iron-carbon alloy of eutectoid composition (0.80% C), including austenite to pearlite and austenite to bainite transformations. Source: *Atlas of Isothermal Transformation and Cooling Transformation Diagrams*, ASM International.

and ended slowly. Plotting the beginning and ending times for many different temperatures resulted in a curve with a c-shape (**FIG. 10.16**). At the highest temperatures, the start and end of transformation were delayed in time. As temperature decreased, the reaction was faster until about 595 °C (1100 °F). At even lower temperatures, the reaction rate decreased again, and a new microstructure formed that was not pearlite. This new phase was named bainite in his honor. These plots are called time-temperature-transformation (TTT) diagrams. Bain and Davenport studied the effect of carbon and manganese on the TTT diagrams.

During his active research career at U.S. Steel Corp., Bain and a number of coauthors published 20 technical papers between 1929 and 1939. He also coauthored a book on tool steels with Marcus Grossman in 1931 and published his book *Functions of the Alloying Elements in Steel* in 1939. In 1935, Bain was promoted by U.S. Steel Corp. to assistant to the vice president of research and technology in New York. He was promoted again in 1943 to vice president of research and technology of the Carnegie-Illinois Steel Corp. and moved to

Pittsburgh, where he later became vice president of research and technology for the entire U.S. Steel Corp. The corporation built a research campus in 1956 in Monroeville, Pennsylvania, which included the Edgar C. Bain Laboratory for Fundamental Research.

He was active in many technical societies. As early as 1925, he served as chairman of the Buffalo Chapter of AIME. In 1937, he served as president of ASM and received Honorary Membership in 1959. He was conferred Honorary Membership by AIME in 1964, and he was made a Fellow of the American Physical Society and the American Society for the Advancement of Science. He was elected into the National Academy of Sciences in 1954.

Bain received many awards for his 20-year career in metals research. They included the Robert W. Hunt Medal in 1929, the Henry Marion Howe Medal in 1931, the Albert Sauveur Achievement Award in 1946, the ASM Gold Medal in 1949, and the John Price Wetherill Medal from the Franklin Institute in 1949. Overseas, he was the first American to receive the Grande Medaille of the Societe Francaise de Metallurgie in France and the Gold Medal of the Japan Institute of Metals.

Edgar C. Bain suffered a stroke in 1959 that left him partially paralyzed. He continued to consult from his home and wrote his final published work, *Pioneering in Steel Research: A Personal Record*, an autobiography, which was published by ASM in 1975 after his death in 1971.

Samuel Leslie Hoyt

Samuel Hoyt was one of the earliest metallurgists to receive his advanced training in Europe. In the first part of the 20th century, the major teaching and research in metallurgy was being done at universities in Germany, France, and England.

Hoyt was a native of Minnesota and graduated from the University of Minnesota in 1909. His main interest at that time was in geology, and he enrolled at Columbia School of Mines in New York to earn his graduate degree. Here he was advised to take Howe's course in metallurgy and a course in metallography, taught by William Campbell under Howe. As with several other students exposed to metallography, Hoyt dropped geology in favor of a degree in metallurgy. After two years at Columbia, he enrolled at the Royal Institute of Technology at Charlottenburg, Germany, to pursue advanced studies under some of the best metallurgy professors in Europe. He took courses under H. Hanemann, W. Guertler, Adolf Martens of martensite fame, and other top European professors. Classmates included Paul Dyer Merica and several other Americans. He took a field trip to steel mills with Merica and a new acquaintance, Guillaume (William) Kroll, the inventor of a process for making titanium. Later in his career, Hoyt returned to Europe to visit laboratories, steel mills, and universities. He spent time with his former teachers, classmates, and other well-known metallurgists in Germany, France, and England.

In 1913, Hoyt returned to Columbia, where he received his doctorate in metallurgy. His first career location was as an instructor at his old school, the University of Minnesota. He taught metallography and the newer field of phase diagrams. Phase diagrams were taught based on J. Willard Gibbs' theory of "heterogenous equilibria," which was first applied to metals by a Dutch metallurgist, H.W. Bakhuis Roozeboom, in 1898. Roozeboom had been largely forgotten, but in his biography, Hoyt said, "Where there had been uncoordinated chaos, the great Dutch scientist brought order and understanding with the most necessary single contribution to physical metallurgy." Hoyt's education in Germany included work on a number of alloy phase diagrams.

After six years teaching, Hoyt was recruited by Zay Jeffries to work at the General Electric Lamp Division, where he worked with Edgar Bain, who had joined General Electric (GE) at the same time. Hoyt conducted research for the next two years on problems with manufacturing tungsten wire for lamps. He transferred to the GE Research Laboratories at Schenectady, New York, in 1921, where he worked under Willis Whitney, the director of research who had built the laboratory for GE. Here he also had contact with Arthur Compton, who had married a college classmate of Hoyt's (**FIG. 10.17**). Compton won the Nobel Prize

FIG. 10.17 Arthur Compton of atomic bomb fame worked with Samuel Hoyt at the GE Central Research Labs.

FIG. 10.18 Irving Langmuir won the Nobel Prize in Chemistry in 1932.

in Physics in 1927 and became famous for his later work on the atomic bomb. Hoyt also worked with Irving Langmuir, a Nobel Prize winner in chemistry. Langmuir had graduated from Columbia School of Mines with a metallurgical engineering degree. (**FIG. 10.18**).

In 1925, GE sent Hoyt to Europe to study developments in tungsten and lamp bulb manufacturing. He used the opportunity to visit teachers and classmates from his time at the Royal Institute a decade earlier. He discussed advances in precipitation-hardened aluminum with his favorite teacher, Professor Guerther, who was working on alloys in the aluminum-zinc system. Alloys in this system with 5% zinc became the major high-strength alloys in the United States after World War II. He also visited a classmate who had been working on tungsten carbide cemented with cobalt, which was used in Germany as a cutting tool superior to high-speed steel. Hoyt brought samples back to GE and, along with Jeffries, helped convince GE to go into production of tungsten carbide tools in a division called Carboloy.

After a decade of working in the most advanced research laboratory in the United States, Hoyt accepted a position in research at A.O. Smith, a manufacturer of steel pipelines, oil field equipment, and chemical pressure vessels in Milwaukee. Working with steel vendors, he became acquainted with metallurgists throughout the industry. He was visited by some of his classmates from

FIG. 10.19 Battelle Memorial Institute in Columbus, Ohio. Source: Wikimedia Commons/Wiki Historian N OH.

Europe and by men he had visited in England. William Kroll also came to see him on his trip to interest U.S. companies in his process for making titanium.

Hoyt accepted a position in 1939 as consultant to the president of Battelle Memorial Institute in Columbus, Ohio, the largest metals research laboratory in the United States (**FIG. 10.19**). As a nonprofit organization, Battelle did contract research for industry and the government. At that time, many companies lacked the facilities and staff to do their own research, and as World War II approached, Battelle worked on many problems with the defense industries.

Hoyt broadened Battelle's activities by using his experience in welding, steel quality, and brittle fracture, especially when all-welded ships began to rupture at low temperatures during World War II (**FIG. 10.20**). Through his contacts at Battelle, he became acquainted with many industrial metallurgists, corporate executives, and government agencies.

At the end of World War II, Hoyt was asked by the army to aid in exploring the metals developments in Germany and the occupied areas of Europe. With his knowledge of the German language and his contacts from universities and from his GE trip in 1925, he was well suited for the job. He saw the devastation of European cities; many laboratories he had previously visited had been moved to temporary sites after their original locations were destroyed during bombing raids.

Upon returning to Battelle, he worked on the institute's plans for expanding into Europe. Two sites had been selected for building laboratories, Frankfurt, Germany, and Geneva, Switzerland. Hoyt traveled to Europe again in 1952 to survey the market in steel research, and he found that many companies had

FIG. 10.20 SS Schenectady, a welded tanker ship that broke in half while docked in 1943.

problems and programs that could not be resolved by their existing staff. He found enough potential research needs that Battelle went forward with its plans.

He worked with Zay Jeffries in 1951 on the World Metallurgical Congress, organized by ASM. He frequently saw Jeffries during his time at Battelle, as Jeffries was a trustee on the board of directors. It had been 30 years since they had worked together at the GE's tungsten wire plant in Cleveland.

Hoyt retired in 1953 and started a consulting practice. He concentrated on his strengths in welding, steel quality, and brittle fracture. The country was just becoming aware of the serious problem of brittle fractures, which destroyed many structures, especially ones constructed by welding. Once a crack started, it could run to destruction, as it did with the all-welded ships during World War II. Adding to the problem was the knowledge that hydrogen in steel and in welds could produce cracks large enough to cause brittle failures.

Most of his work involved large structures, such as dam gates and penstocks in hydroelectric power plants, where welding of heavy plate was both a steel quality issue and a welding problem. Later, Hoyt became involved in brittle failures in rocket motor cases. These failures, and the resulting destruction of intercontinental ballistic missiles upon launching, set back U.S. defense efforts by several years. It was brought under control after research was conducted into steel cleanliness, improved welding techniques, new tools for nondestructive testing to locate defects in the interior of metals, and a new approach to fracture toughness technology.

In 1959, Hoyt was again recruited to travel to Europe for his fourth visit since his college days. He was representing the North Atlantic Treaty Organization (NATO) in studying the metals industry of Europe. This time he took his wife Edyth with him. He visited plants, research laboratories, and universities during the day, and they enjoyed concerts, plays, museums, and art galleries in the evenings and weekends. By this time, Hoyt was well acquainted with Europeans and was very conversant in the German language. He continued consulting throughout the 1960s and finished his autobiography, *Men of Metals*, published in 1979 by ASM (**FIG. 10.21**). In it, he described his career and all the men he knew, from European pioneers such as Henri Le Chatelier, William Guerther, and Adolf Martens to those he knew from teaching, working at GE, the steel industry, research advising, and consulting. His book contains more than 300 names of metals men he knew during his career. As he describes them, "They were the leading metallurgists during the first half of the 20th century, the time when the science of physical metallurgy developed," (Ref 2).

Francis L. VerSnyder

The invention of the single-crystal jet engine blade was one of the 50 greatest advances in metallurgical history. It was an invention unknown to most Americans, yet it has advanced the technology of air travel since the 1970s.

FIG. 10.21 Samuel Leslie Hoyt and his book, *Men of Metals*.

The research program started in the early 1960s at Pratt & Whitney, a subsidiary of United Technologies Corporation. Frank VerSnyder, who had just joined the company, headed the research. Pratt & Whitney had expanded their research effort to make major improvements in the metal alloys for the hottest part of the jet engine. It was a long-range project whose goal was to eliminate failures that sometimes occurred at grain boundaries of cast turbine blades. These blades were made of nickel-base alloys that withstood the extreme temperatures in operation.

VerSnyder grew up in Watertown, New York, as an only child in a single-parent household (**FIG. 10.22**). His mother died at his birth, and his father was much older. He attended local parochial schools through high school and joined the army in 1943 at age 18.

He served in the 54th Armored Infantry Division in Italy, France, and Germany. He was wounded several times and received the Purple Heart with oak leaf cluster and campaign ribbons with four battle stars. VerSnyder was discharged in 1945, still only 20 years old.

He attended the University of Notre Dame on the G.I. Bill, receiving a Bachelor of Science degree in metallurgical engineering in 1950. He joined the small aircraft engine division of General Electric in Lynn, Massachusetts. By 1955, he transferred to the General Electric Research Laboratory in Schenectady, New

FIG. 10.22 Francis Louis VerSnyder, the leader of a group of scientists and engineers who invented single-crystal jet engine blades.

FIG. 10.23 Maurice "Bud" Shank recruited a team of scientists and engineers to improve jet engine performance.

York, where he remained until 1961. He was recruited by Maurice Shank, director of the research laboratories at Pratt & Whitney, to head a new group that had a 10-year goal of improving the reliability of engine blades operating in the hottest part of the engine (**FIG. 10.23**).

When molten metal is cast into a ceramic mold, solidification starts at the cold wall of the mold. The solid metal grows as single crystals into the remaining liquid until the complete space of the mold is filled. Each of these growing crystals (called grains) contacts other grains that are growing at different orientations in space. Where they meet, they form boundaries that must adjust to these different orientations. These boundaries are weak regions in the alloy.

The cast turbine blades had many grains that could be seen by chemically etching the surface. Some areas of boundaries lay along the length direction of the blade; others, called transverse boundaries, were across the blade. In operation, the blades spin at very high revolutions per minute, so the major stress was along the length of the blade. The boundaries that cross the blades were the ones that were weak during operation. Failure occurred by separation of these boundaries.

The original goal of the program was to grow grains in only one direction, the length of the engine blade, thus eliminating the grain boundaries across the blade. To accomplish this goal, the beginning of solidification had to occur from only the bottom of the mold. This required heating the mold to high temperature to prevent solidification on the mold walls. Solidification then was limited to a water-cooled plate at the bottom of the mold. Even though many grains formed, they grew only in one direction as the heat was withdrawn through the bottom. The process required precise control of molten metal temperature, mold shape, wall thickness, mold temperature, and withdrawal of the mold as the liquid/solid interface moved up the mold, which ensured that the newly formed solid grains did not re-melt.

After this process was completed in the laboratory, it was translated into production by the jet engine blade vendors. The first engine to use the directional solidified (DS) blades was the F-111 fighter plane (**FIG. 10.24**). During the 1970s, DS blades were used in the Pratt & Whitney engines for the Boeing 747 and the McDonnell Douglas DC-10, and they were also used in the new fighter planes, the F-15 and F-16. Engines built by General Electric and Rolls-Royce also used DS blades.

With the success of laboratory work on DS blades, attention turned to making a single-crystal blade. This was the ultimate accomplishment in the science of metallurgy. Single crystals had been made and studied in the past, but they were small samples and were only in pure metals—copper and zinc. No attempt had ever been made to make a single crystal in a metal as complicated as the

FIG. 10.24 The F-111 fighter plane was the first to use directionally solidified jet engine blades.

nickel-base alloys used at the extreme temperatures of the jet engine, and no one had ever contemplated making a useful part with it. This was virgin territory for metals technology.

The first breakthrough came when a member of the team, Steven Copley, was making a coiled spring to study the properties of DS crystals. He used a mold in the shape of a coiled spring suspended in the liquid metal with the DS crystals growing from the bottom of the regular water-cooled plate. As the growing crystals entered the opening of the coiled mold and grew up the spiral shape, they had various orientations in space. Crystal orientations grew at different rates into the liquid metal, with one orientation growing faster than the others. As they grew up the spring mold, the favored orientation grew ahead of the others until only a single orientation was present, a single crystal (**FIG. 10.25**). This spiral mold, called the "pig tail," became the process for producing single-crystal jet engine blades.

The single-crystal blade revolutionized the operation of jet engines and aircraft travel. The engines operated at temperatures of 35 to 65 °C (100 to 150 °F) higher than before. They improved the efficiency of operation, fuel economy,

FIG. 10.25 Three configurations of grains for the jet engine blades.

FIG. 10.26 The Boeing 747 used single-crystal jet engine blades to replace the directionally solidified blades. Source: Boeing.

and time between engine overhauls. These higher temperatures required new alloys and new ceramic coatings to protect the blades. All through the program, alloy composition was adjusted to account for different grain formations and solid solution strengthening. The final alloy had the following composition: 10% chromium, 5% cobalt, 4% tungsten, 1.5% titanium, 5% aluminum, and 12% tantalum. By the end of the 1970s, single-crystal blades replaced the DS blades in the Boeing 747 and were also used in the new Boeing 757 and 767 and the Airbus A310 (**FIG. 10.26**). Single-crystal blades are now standard for all big engines for commercial and military aircraft.

Frank VerSnyder earned his first award, the Henry Marion Howe Medal from ASM, for his work on the microstructure of jet engine alloys. It was 1954, and he had been in research only four years. For his inventions in DS and single-crystal engine blades, he received the George Mead Medal for Engineering Achievement from United Technologies in 1965, the Clamer Award from the Franklin Institute in 1973, and the ASM Engineering Achievement Award in 1975. He was selected for the National Academy of Sciences, Engineering Division, in 1981. Frank VerSnyder received the highest award given a civilian citizen, the National Medal of Technology and Innovation, presented in the Oval Office by President Reagan in 1986 (**FIG. 10.27**).

His research into the development of DS and single-crystal casting for jet engines is listed by AIME as one of the 50 greatest advances in the history of metals. He continued research at the Pratt & Whitney engine laboratories for the rest of his career. He died at the age of 64 in 1989.

FIG. 10.27 Frank VerSnyder received the National Medal of Technology and Invention in 1986.

REFERENCES

1. H.M. Howe, *The Metallurgy of Steel*, Vol 1, The Scientific Publishing Company, 1890
2. S.L. Hoyt, *Men of Metals: An Exciting Career among the Trailblazers of Modern Metallurgy*, American Society for Metals, 1979
3. E.C. Bain, *Pioneering in Steel Research: A Personal Record*. American Society for Metals, 1975

Selected References

J.B. Austen, *Edgar C. Bain*, National Academy of Science, Washington DC, 1978. Retrieved from www.nasonline.org/publications/biographical-memoirs/memoir-pdfs/bain-edgar-c.pdf

J.B. Austin, Henry Marion Howe, *The Howe Memorial Lecture*, presented at the Metal Society of American Institute of Metallurgical Engineers, Iron and Steel Division (Englewood CO), 1970. Published as J.B. Austin, Henry Marion Howe, *Metallurgical and Materials Transactions B*, Vol 1 (No. 7), July 1970, p 1795–1803. Retrieved from https://link.springer.com/article/10.1007/BF02642773

G. Burgess, *Henry Marion Howe*, Vol 21, *Memoirs of the National Academy of Sciences*, National Academy of Science, 1927. Retrieved from www.nasonline.org/publications/biographical-memoirs/memoir-pdfs/howe-h-m.pdf

R.A. Daly, *Albert Sauveur*, Vol 22, *Memoirs of the National Academy of Sciences*, National Academy of Science, 1940. Retrieved from www.nasonline.org /publications/biographical-memoirs/memoir-pdfs/sauveur-albert.pdf

A. Giamei, Development of Single Crystal Superalloys: A Brief History, *Advanced Materials and Processes*, Vol 171 (No. 9), 2013, p 26–30

Z. Jeffries, *Paul Dwyer Merica*, National Academy of Science, 1959. Retrieved from www.nasonline.org/publications/biographical-memoirs/memoir-pdfs/merica-paul.pdf

P.D. Merica, R.G. Waltenberg, J.R. Freeman, *Constitution and Metallography of Aluminum and Its Light Alloys with Copper and with Magnesium*, Nabu Public Domain Reprints, 1919

P.D. Merica, and R.G. Waltenberg, *Heat Treatment of Duralumin*, Nabu Public Domain Reprints, 1919

W.D. Mogerman, *Zay Jeffries*, American Society for Metals, 1973

W.D. Nix, *Zay Jeffries*, National Academy of Science, 2013. Retrieved from www.nasonline.org/publications/biographical-memoirs/memoir-pdfs/jeffries-zay.pdf

M. Shank, *Frank Louis VerSnyder*, Vol 4, *Memorial Tributes: National Academy of Engineering*, National Academy Press, 1991. Retrieved from www.nap.edu /read/1760/chapter/54

Chapter Eleven

The Integrated Steel Industry—1901 to 1959

The United States Steel Corporation

The birth of the big integrated steel industry began with the formation of the United States Steel Corporation (USS) in 1901. *The Corporation*, as it was known for the next century, was initially composed of the Carnegie Steel Company, its biggest producer, and the Federal Steel Company, its number two producer. Carnegie Steel was well known as the leading steel company in the Bessemer Steel Age from 1875 to 1900. Federal Steel was an unknown company assembled by a lawyer and judge from Illinois named Elbert H. Gary, who had combined nearly 40 plants that made wire. This group was financed by J.P. Morgan (**FIG. 11.1**). Judge Gary became the first chairman of USS in 1902 and Charles Schwab of Carnegie Steel was the first president.

The new USS owned 213 steel mills, 41 iron mines, 112 ore boats, and 57,000 acres of coalmines. The company controlled six million tons of production, about half of the nation's total. It was capitalized at $1.4 billion ($40 billion in 2015 dollars), and shares were offered for sale to the public. In the first full year of operations, they had revenue of $560 million and earnings of $34 million.

After a year as president, Schwab was forced out of the company over differences with Gary (**FIG. 11.2**). Schwab was used to having free reign at Carnegie Steel and did not fit in with the company's corporate image under Gary. Schwab would later become president of Bethlehem Steel, which he would lead to become a major competitor of USS.

FIG. 11.1 John Pierpont Morgan, the financier who assembled many companies to form United States Steel Corporation, circa 1918.

FIG. 11.2 Judge Elbert Henry Gary, the first chairman of U.S. Steel, circa 1915.

Many new steel companies were formed at this time to become competitors to USS. They included Youngstown Sheet and Tube Company of Youngstown, Ohio; American Rolling Mill in Middletown, Ohio; and Inland Steel Company in Chicago. With this new competition, USS acquired other mills, adding approximately 15% to its capacity.

In 1906, with most of their steel mills located in the east, The Corporation under Judge Gary planned to build a new mill at a large location in Indiana, bordering on Lake Michigan. This isolated location required that The Corporation build the infrastructure needed for an integrated steel mill, with railroads, a harbor for receiving raw materials, and a new town for workers and service personnel. The project was enormous for the time, and this steel mill would be among the largest in the world. By 1911, the city, named Gary in honor of the chairman, had a hotel, churches, a hospital, and homes for a population of nearly 20,000. The steel plant housed eight blast furnaces, 47 open-hearth furnaces, hundreds of coke ovens, and rolling mills to produce railroad rails, axles, and bars. Later, sheet and plate mills were installed for the new auto and home appliance industries. As Brian Apelt reported in *The Corporation*, "Gary was a gem, a symbol of the bright new century and the things it seemed to promise," (Ref 1).

As The Corporation grew, it relied on immigrants for labor, just as Andrew Carnegie had in the 1880s and 1890s. In 1910, half of its workers were from Poland or the Slavic countries. During World War I, the supply of immigrants was closed, and recruits from the South filled the labor needs. There were 3,000 recruits in Gary and 4,000 in Pittsburgh by 1919. With the war over, wages and working conditions led to strikes in many plants. At Gary, the Indiana State Guard was called to restore order.

The Corporation experienced successful years during the 1920s. The increased production in manufacturing autos, farm and construction machinery, high-rise buildings, and bridges created a need for additional capacity. They built sheet mills in Indiana, Pennsylvania, Ohio, and Alabama. They also built pipe mills to supply the ever-increasing production of oil. During this 10-year period, total earnings reached $1 billion, with two-thirds of that paid in dividends. All of this success was accomplished under Judge Gary, who continued to work until he was 80 years old.

The financial success of the 1920s obscured the fact that the company was falling behind in new technical developments. A small competitor, the American Rolling Mill Company in Hamilton, Ohio, developed a continuous sheet rolling process in 1923, replacing the old hand sheet rolling technique. This was a major breakthrough, not only in the production of sheet but also in supplying long lengths for stamping presses, enabling them to be more efficient and to make bigger parts, especially for autos. There were seven continuous sheet rolling mills in production by 1929. By the time USS installed its first mill at the Gary Works, there were 12 other mills in operation. Years later, they were still working to catch up with this most important advance in producing sheet, which was becoming a major product in steelmaking. The Corporation was slow to recognize that it was no longer the industry leader in technology.

With the depression of the 1930s, steel demand plummeted. The Corporation produced 14 million tons of steel during the peak year of 1929. In 1933, they only produced 3.8 million tons. The loss that year was $91 million, and they had only 80,000 workers, compared with the 240,000 workers during the 1920s. At the lowest point in 1933, plants were operating at 9% capacity. Production increased as government programs began construction under the Works Progress Administration (WPA), and auto sales increased during the 1930s.

In 1932, USS appointed a new chairman to guide them through the coming depression, Myron Taylor (**FIG. 11.3**). Taylor was a lawyer with previous experience in restructuring big corporations. He joined USS in 1925 as a director on the board of trustees and chairman of the finance committee. He was familiar with the company and its many management and production problems by the time he was promoted.

Taylor's first initiative was to merge the two biggest divisions into one unit. Carnegie Steel and Illinois Steel represented three quarters of the total production capacity; they were merged and renamed Carnegie-Illinois Steel Corp.

FIG. 11.3 Myron Charles Taylor, the second chairman of U.S. Steel, who led the company through the Great Depression.

Taylor hired an outside executive named Benjamin Fairless from Republic Steel Corp. to be the president of the new organization. He also promoted a long-time executive, William Irving, to vice president of the corporation. Irving later became president.

Taylor had formed a committee of engineers when he was chairman of the finance committee, where tasked with the responsibility of studying the entire corporation to improve efficiency and to identify obsolete or outmoded plants. In 1935, he added to this committee by hiring an outside consulting firm of industrial engineers to examine every operation in all of the facilities, from raw materials to finishing mills.

With improved business conditions in the late 1930s, production had increased from 43% of capacity in 1938 to 63% in 1939. Taylor tried to implement many of the recommendations of these two consulting groups, but the sheer size of the effort and the beginning of production for World War II delayed the process. One major proposal—for a board to oversee long-range planning—was never implemented, and this deficiency was reflected in the postwar period.

Under Taylor's leadership, The Corporation closed production in obsolete plants equal to five million tons of annual capacity, or a third of the company total. Some of these plants were still using Bessemer converters, which were now 30 years out of date. However, Taylor added six million tons of capacity in new plants, with two thirds of that for continuous rolling mills producing sheet and strip products.

By 1940, the defense effort to rebuild the U.S. armed forces increased, and the United States was also helping England and France through the lend-lease program. During the five years leading up to Japan's surrender, the United States produced $190 billion in armament, with a third of all steel made by USS. For example, a USS-owned mill in Lorain, Ohio, produced over 1200 miles of 24-inch-diameter pipe for the big pipeline carrying oil from Texas to New Jersey. The Corporation built a new, government-funded steel mill at Geneva, Utah, to supply steel for the West Coast shipbuilding industry. They bought this mill after the war for $40 million. Throughout the war, the USS American Bridge Division erected government-funded factories and buildings for defense production. The Corporation's shipyards built 911 ships during the war, including 119 tank landing ships and 22% of all destroyers.

The postwar era saw the cancellation of nearly all government contracts, producing severe unemployment. With the conversion back to consumer goods and the introduction of new technologies developed during the war, the economy soon revived, with demand for steel exceeding wartime levels. The 1950s, which was the most prosperous decade since the 1920s, began with the Korean conflict, which again increased demand for armament. Next came the Cold War and the continued need for a strong defense industry. New technologies such as the nuclear-powered Navy and the beginning of the nuclear power energy industry increased demand for steel. The auto industry was building more and heavier cars, and appliances became household necessities. In addition, for the first time since construction of the Empire State Building in 1931, a high-rise building was erected—the United Nations headquarters in New York (**FIG. 11.4**). The 1950s would see the beginning of a boom in such buildings, and the Federal-Aid Highway Act of 1956 (to build the interstate highway system) added to steel demand as well.

With forecasts for increased use of steel, USS started a program of modernizing its older mills and adding equipment for new products. The most ambitious project was a new steel works at a rural site on the Delaware River in

FIG. 11.4 United Nations Headquarters, the first high-rise built after the Great Depression. U.S. Steel supplied the steel and completed the construction work.

Pennsylvania, at the New Jersey border. This works required rail line infrastructure, a harbor for unloading ore ships, blast furnaces, open-hearth furnaces, coke ovens, and rolling mills for various products. The iron ore for this mill was imported from a deposit owned by USS in Venezuela. This plant was the largest ever built in the United States on a new site. A new town was built several miles away to house workers and all the related services. The plant was named the Fairless Works and the town Fairless Hills, after Benjamin Fairless, the new chairman of The Corporation.

The Corporation built a new headquarters in Pittsburgh in 1953. It was a 41-story building made of steel and stone (**FIG. 11.5**). In 1956, they built a new research and development campus at Monroeville, near Pittsburgh. USS had first established a central research laboratory in 1928 at Kearny, New Jersey, in an unused building owned by the Federal Shipbuilding and Drydock Company,

FIG. 11.5 Headquarters of The Corporation, erected in Pittsburgh in 1953.

a division of the company. One of the first research employees was Edgar Bain, who became world famous for his studies on the transformation of austenite to pearlite at constant subcritical temperatures. A fundamental research laboratory on the new campus was named in his honor.

The 1950s was very profitable for the American steel industry; it ended with the understanding that there was a great future for steel, and it would be led by The Corporation. However, in the last months of 1959, negotiations between the major integrated steel industry and the United Steel Workers Union ended with a four-month strike that eventually changed the industry beyond recognition.

Bethlehem Steel Corporation

Bethlehem Steel Corporation debuted as Bethlehem Iron Company in 1860 with blast furnaces and rolling mills for wrought iron railroad rails. The company hired John Fritz from Cambria Iron Company in Johnstown, Pennsylvania, to be general manager (**FIG. 11.6**). After making iron rails for several years, Bethlehem

FIG. 11.6 John Fritz was the manager of Bethlehem Steel for over 30 years. He remade the company from a minor rail producer into a major supplier of heavy forgings and armor plate. Source: Linderman Library, Lehigh University.

contracted with Alexander Holley to build one of the first Bessemer plants to convert to steel. Bethlehem installed four Bessemer converters and new rolling mills for steel rails. The plant went onstream in 1873 with an annual capacity of 35,000 tons and a reputation as the finest plant in the United States. However, it was a small plant, and its eastern location was problematic, as rail use was moving west.

Under Fritz's leadership, the company built facilities for making large forgings and heavy plate, including armor plate for the Navy. Previously, these products had been imported from England and Germany due to a lack of domestic manufacturers. Bethlehem Steel was soon recognized as a producer of ship propeller shafts, large gun barrels, and heavy battleship armor equal to any provided from overseas.

The principals put the company up for sale after the turn of the 19th century. Charles Schwab bought 160,000 shares, giving him controlling interest while he was still president of USS Corp. (**FIG. 11.7**). When he was forced out of The Corporation, he turned his full attention to developing Bethlehem Steel from

FIG. 11.7 Charles Schwab, past president of Carnegie Steel and U.S. Steel, became the major stockholder and chairman of Bethlehem Steel. Source: www.explorepahistory.com, image donated by Corbis-Bettmann.

a minor steelmaker into a competitor to his former company. In 1913, Schwab purchased a shipyard in Massachusetts that specialized in naval warships including submarines.

When World War I started in 1914, the federal government passed a Neutrality Act to keep the United States from becoming involved in the conflict. This prevented trade in armaments with England, France, and Russia. Schwab ignored the Act, as did many other companies, and received orders for gun barrels, armor plate, and ammunition from all three allies. Bethlehem Steel opened a munitions plant to supply overseas trade, and over the next four years, as the war continued, it became the largest plant manufacturing artillery shells in the world, producing 750,000 shells per month. In addition, it produced 60% of all finished guns built for the United States.

Steel shipments during the war period were 6 million tons, and employment increased from 15,000 to 94,000 workers. The shipyard received orders for freighters and destroyers, significantly adding to profits. Earnings in 1916 were \$62 million, with Schwab receiving \$2.4 million in dividends and new stock worth \$17 million. With these wartime profits, Bethlehem purchased the Pennsylvania Steel Company for \$32 million. They acquired the Steelton plant, built for Edgar Thomson and Tom Scott during the Bessemer Steel Age, and a newer deep-water plant at Sparrows Point, Maryland. Bethlehem now had 2.2 million tons of annual capacity, exceeding the capacity of all competitors except USS. In the last two years of the war, Bethlehem's earned income was \$111 million.

Bethlehem used its wartime profits to acquire other steel companies. In 1922, the company purchased the Lackawanna Steel Company outside Buffalo, New York, for \$54 million, raising its annual capacity to 5.5 million tons. A year later it bought the Midvale Steel Company for \$95 million, which included Cambria Steel, where John Fritz had invented the three-high rolling mill and William Kelly had worked on his air blowing process. Bethlehem renamed Cambria Steel the Johnstown Mill. The addition of both Midvale and Johnstown increased Bethlehem's annual capacity to 10 million tons.

Integrating these various steel operations to improve efficiency and lower overhead costs proved to be a greater challenge than expected. As quoted by Kenneth Warren in his book *Bethlehem Steel*, "Much of the effort of Bethlehem Steel through and beyond the 1920s would be devoted to that task," (Ref 2). In addition to the Sparrows Point Plant, where they modernized and added capacity to become one of the largest plants in the country, they had acquired older plants in need of significant maintenance and upgrading.

Profits during the 1920s were good, although Bethlehem's product mix was mostly heavy forging and plate, with less sheet and other light products. This turned out to be a handicap in return on assets. Also, with most of their plants except Lackawanna located to serve eastern markets, they were missing the growing needs in the Midwest. In trying to solve these problems, Bethlehem made an attempt to purchase the Youngstown Sheet and Tube Company (YS&T) in Youngstown, Ohio. It was the third largest steel company in the country, with 3 million tons of capacity and plants in Youngstown and South Chicago. The purchase of YS&T gave Bethlehem assets of over $1 billion, putting them in range of USS, General Motors, and Standard Oil Company of New Jersey, a goal that had been an obsession with Schwab ever since he became the major stockholder of the company. The merger with YS&T was approved by the board and a vote of the stockholders, but the closing was opposed by a minority of the shareholders, led by Cyrus Eaton, and this group went to court to stop the purchase.

During the subsequent trial, it was revealed that Bethlehem had a secret bonus plan that distributed large amounts of money to Eugene Grace and the other top executives (**FIG. 11.8**). In 1929, Grace was forced to reveal his income, which was $1.6 million, larger than any other executive in the country. With this information, another group of shareholders took Bethlehem to court to force

FIG. 11.8 Eugene Grace was president and then chairman of Bethlehem Steel from 1916 until 1959. Source: www.explorepahistory.com, courtesy of the NYPL digital gallery, New York Public Library.

them to provide information about the bonus plan that Schwab had secretly controlled for years without revealing it to the stockholders. At the trial, they found that for a period in the 1920s when dividends were $25 million, the top executives received bonuses of $14 million. The press and Wall Street condemned the secret bonus plan, both for the extreme bonuses and for the company's failure to reveal this information to the shareholders and to the public.

As industry collapsed during the Great Depression, steel shipments reached lows not seen during the previous 20 years. In 1932, Bethlehem operated at just 20% of capacity, with a loss of $19 million. A major problem for them was the location of mills and the high percentage of heavy products. When the recovery started in 1933 and beyond, the increased business was in lighter flat-rolled products for the automobile and appliance markets in the Midwest. Bethlehem only had one mill that could reach this market, the Lackawanna plant. They built two new hot continuous sheet rolling mills starting in 1935, one in Lackawanna and the other at Sparrows Point. Both mills had annual capacities of 600,000 tons. This was a beginning in producing light sheet and plate, but Bethlehem was still primarily a producer of heavy products. One large project for Bethlehem was building the Golden Gate Bridge, completed in 1937. It required 100,000 tons of steel, which was produced at the Sparrows Point plant.

With the start of World War II in Europe, Bethlehem received $300 million in orders from England. It was production for overseas orders in World War I that had allowed Bethlehem to accumulate excessive profits; this time, the orders went through the War Production Board instead of J.P. Morgan in London, thereby limiting profits.

Bethlehem's war production was summed up by Kenneth Warren in his book *Bethlehem Steel*: "Over the next few years (WWII) though steel production reached new peaks, Bethlehem's greater glory was in its shipyards, its most vital contribution to the war effort," (Ref 2). Shipbuilding employment increased from 7,000 to 182,000, accounting for 95% of employment growth for the entire corporation. The Fore River Shipyard in Quincy, Massachusetts, built four aircraft carriers, a battleship, and 31 cruisers (**FIG. 11.9**). In four years, Bethlehem built 1127 ships and produced 73 million tons of steel. Seventeen million tons were made in the Sparrows Point plant. Bethlehem furnished the Navy with one-third of its needs for armor plate and gun forgings.

The 1950s were very productive and profitable for the steel industry, as demand for steel exceeded that of World War II. At times, Bethlehem operated at 100% capacity, with an average of 150,000 employees. Total net income for the period 1950 to 1959 was $1.37 billion. With these reserves, the company added 5 million tons of steel to its annual capacity, bringing the total to 20 million tons.

FIG. 11.9 The USS Massachusetts was built at the Bethlehem Fore River Shipyard and commissioned in 1942.

In 1954, Bethlehem attempted to acquire Youngstown Sheet and Tube to gain additional markets in sheet products in the Midwest. However, the merger was prevented by the Justice Department on the grounds that it would violate the Clayton Antitrust Act. Thus, Bethlehem was again denied the opportunity to increase its reach into the auto and appliance producers in Ohio, Illinois, and Michigan, with only their light production limited to Lackawanna.

In 1956, Bethlehem announced a program to expand production to 23 million annual tons. Two million would be at Sparrows Point, increasing that plant to eight million tons with a tenth blast furnace and a new open hearth shop. When expansion was completed in 1958, Sparrows Point was the largest steel mill in the world. All of this optimism was based on steel consumption continuing well into the next decade or two.

The industry increased prices 6% in 1957, despite the company's earnings—the highest in modern times. At the same time, Bethlehem distributed the largest bonuses to executives since 1929. Grace was again the highest paid executive in the United States, and of the 18 highest paid corporate officers in the country, 11 were at Bethlehem. This insensitivity in the matter of the bonuses, and the inflation the industry was creating, drew the attention of the unions, the press, and the federal government.

In early 1959, the steel industry failed to understand that the high production and profits they had enjoyed for 10 years had come at an unsustainable cost. While they were building new capacity with the open hearth process that had dominated steelmaking for over 50 years, new technology was coming onstream in Europe and Japan. First was the basic oxygen furnace (BOF), which used oxygen blown through the heat of molten metal to convert it to steel. This novel process could convert cast iron from the blast furnaces to steel in 45 minutes compared with 8 hours in the open hearth. The process was invented in Austria and was installed in other countries as new open hearths were erected at Sparrows Point. One small U.S. producer, McLouth Steel Company, had built a 35-ton unit in 1954 (**FIG. 11.10**). Jones and Laughlin Steel Company followed by building several in 1959. This left the two biggest steel companies in the United States with new open hearth shops that would become obsolete in the 1960s.

The second new technology that was developing overseas was the continuous casting machine. Molten metal from the steelmaking furnaces was ladled

FIG. 11.10 McLouth Steel Company built the first basic oxygen furnace in the United States. This process would eventually replace all open hearths. Source: Wikimedia Commons/Jmk7.

into a machine that allowed solidification as the metal moved through the unit and a semifinished product left the machine. This process eliminated the old system of casting molten metal into large ingots that had to be reheated and reduced to semifinished products by forging or in a rolling mill.

Little Steel Companies

Steel companies that emerged in the Midwest to serve the growing manufacturing markets in Ohio, Illinois, and Michigan were much smaller than United States Steel Corporation (USS), and were therefore referred to as "Little Steel." One of the first in production was Youngstown Sheet and Tube Company (YS&T) in Youngstown, Ohio. It was formed in 1900, with George D. Wick as president. He died in 1912 with the sinking of the RMS *Titanic*. In 1923, YS&T purchased the Steel and Tube Company of America with plants in East Chicago and Indiana Harbor, Indiana.

Another Little Steel company in the Midwest was Inland Steel, based in Chicago. It was formed in 1901 with an open hearth shop in East Chicago, Indiana. Production increased, and by World War I it reached one million tons annually. In 1932 during the Great Depression, Inland's net loss was only $3 million. They recovered to a $9 million profit in 1935, the second highest in the industry. Jones and Laughlin Steel Company (J&L) was an old name in the Pittsburgh area. It began steel production with a Bessemer shop in 1886.

The American Rolling Mill Company (Armco) started production in 1899 in Middletown, Ohio. A new plant was added to production in Ashland, Kentucky, in 1901. In 1923, the plant developed a major advance in steelmaking with the hot continuous rolling mill for sheet. This process revolutionized sheet making for the auto and appliance industries, and the Little Steel companies would be the first to install the rolling mill.

Another early steel company was Republic Iron and Steel Company. It started in 1899 by assembling small companies in Ohio, Alabama, and Chicago. Cyrus Eaton of Cleveland acquired Republic in 1927, added other mills, and changed the name to Republic Steel Corp. Eaton hired Tom Girdler from J&L Steel Company to be president and he continued to add other small mills to Republic. Eventually it became the third largest steel company in the country.

The National Steel Corporation was formed with Weirton Steel and various small steel companies in the Midwest in 1929. It was immediately successful due to its locations near the auto steel markets and its emphasis on sheet products. It was the only steel company to earn profits during the Great Depression. From 1930 to 1935, it earned $26 million, and in 1935, its net income of $11 million was higher than that of USS ($1.1) and Bethlehem ($4.3) combined.

Labor

Labor unrest was common in the steel industry. Low wages, a hot and dirty work environment, unsafe conditions, and poor treatment of workers led to strikes. One serious strike occurred at YS&T in 1916. Workers were fired on by company guards, killing two and wounding more than 100. The incident resulted in a riot that burned part of the city of Youngstown. These isolated strikes were often limited to one company or one plant and were therefore never successful. Steelworkers were unable to form an organization that would lead to industrywide action, and there was no legal basis for workers to join a union like those that protected corporations. Therefore, company officials could call the police and even the National Guard to break up a strike and protect their property. This lack of legal protection for workers was corrected during the Franklin D. Roosevelt administration.

Congress passed a labor bill written by Senator Robert F. Wagner of New York that became known as the National Labor Relations Act, signed by President Roosevelt on July 5, 1935 (**FIG. 11.11**). This Act gave workers the right to organize and bargain with management for wages, hours, and other benefits. The Act also established a National Labor Relations Board (NLRB) to oversee elections and to examine company labor policies. Forty-five years after the

FIG. 11.11 Robert F. Wagner was responsible for writing the National Labor Relations Act.

worker struggle at the Homestead Plant of Carnegie Company, where Henry Frick faced down strikers with 300 armed guards (see Chapter 3, "The Age of Steel—1870 to 1900," in this book), labor finally had the backing of the federal government for the right to form unions to negotiate with management.

However, the immediate problem was finding an organization that could assist steelworkers with organizing a nationwide union movement. The natural candidate was the American Federation of Labor (AFL), the union of skilled craftsmen, which already had contracts with the steel industry. However, the AFL was more interested in protecting its own workers than organizing unskilled workers. John L. Lewis, president of the national United Mine Workers union for coal miners, voiced his support for steel workers at the annual meeting of the AFL without success. In 1936, he broke away from the craft union to form a Steel Workers Organizing Committee (SWOC). Using officials from his own union to head the activity, he recruited workers from the steel industry. The SWOC demanded a $5 day, 40-hour work week, plus time and a half for overtime.

Lewis's first goal was to organize the largest steel companies, believing that the smaller firms would follow. His breakthrough came in March 1937 when he and Myron Taylor, president of USS, signed an agreement to recognize the SWOC for its 225,000 workers. However, Lewis had misjudged the reaction of the Little Steel companies and Bethlehem Steel. He faced two of the most anti-union men in the history of steel, Eugene Grace of Bethlehem and Tom Girdler of Republic.

Eugene Grace began work as a crane operator at Bethlehem in 1898 upon graduation from Lehigh University. Under Charles Schwab, he advanced to president within 16 years. Grace adopted his attitude toward labor from Schwab, who had been superintendent of the Homestead plant during the 1892 strike. From that event, Schwab formed his extremely negative attitude toward unions.

Tom Girdler was another enemy of the labor unions (**FIG. 11.12**). He became president at J&L Steel Company, but left to take over the newly formed Republic Steel Company in 1930. To protect their companies during the unrest of the 1930s, Bethlehem and Republic Steel, along with other Little Steel companies, began stockpiling guns, ammunition, clubs, and tear gas. Management's contentious attitude toward labor had not changed since the Homestead strike.

However, after the Wagner Act, which legalized recruitment of union members and encouraged elections, the companies ignored the law, rejecting workers organizing just as they had in the past. The SWOC asked for a meeting with Grace and Girdler to discuss holding elections to determine if the majority of their workers wanted a union or not. Girdler's answer was to close the plant in Canton and lay off the workers. Eventually, Grace and Girdler met with Philip Murray, and they agreed among themselves to bargain over wages and hours,

FIG. 11.12 Tom Girdler, president of Republic Steel, testifying about the company's actions during the Little Steel strike of spring 1937.

but to refuse to hold elections or sign to recognize a union. Grace, Girdler, and Murray were all in violation of the Wagner Act.

The SWOC was prepared to act at the Republic plant in South Chicago. There were already 33,000 steelworkers on strike at other Little Steel plants. The company had prepared by trucking in food and cots to house the replacement workers—supervisors, managers, and imported stand-ins for the strikers. Republic intended to keep operating during the strike.

During the last week of May 1937, the SWOC members attempted to demonstrate at the gates in front of the plant, and each time they were turned away by the police. On Sunday, May 30, 1937, the men made a new attempt to reach the gates. It was Memorial Day, and wives and children joined them, along with workers from other steel plants and volunteers from the public. They were met by a line of 200 police preventing them from reaching the gates. Strike leaders tried to bargain with the police to let them through for a peaceful picket line, but tensions escalated. As the crowd began to press forward, they were met with tear gas and gunshots into the air. As people in the back kept pushing forward, the police responded by throwing more tear gas and firing into the crowd (**FIG. 11.13**). Everyone turned to escape as the gunfire continued along with beatings by police clubs.

Ten men were killed and over 100 people needed medical attention. Afterward, the police, newspapers, and Republic Steel blamed the workers for the massacre. Later investigations by Congress found that all the dead were shot in the back or side while trying to flee the area. It was Congress that finally published the real facts of what happened during the Memorial Day Massacre.

The strike failed and workers returned to their jobs. Many were fired for their union activities, and some were blackballed, making them unable to find work in their communities. But the SWOC did not accept the situation. They turned to the federal government and the courts for help. In the spring of 1938, the NLRB found that Republic Steel had violated the Wagner Act. It ordered the company to rehire thousands of workers with back pay and to award payment to the injured and the families of those killed. The company appealed the order, but the court ruled in favor of the NLRB.

During the next several years, the union added enough workers to reach its goal of more than 50% membership. The NLRB supervised the election and certified that the company had to sign a contract with the union. Republic

FIG. 11.13 The Memorial Day Massacre of 1937 killed 10 men and wounded more than 100.

appealed to the courts again, but in 1942, the Supreme Court handed down the decision that Republic was in violation of the Wagner Act. The Little Steel industry finally had to give up its fight to avoid union shops. By then, with World War II underway, the federal government refused to let contracts for defense material go to companies without a union contract. The workers won, but at a price they would never forget. The whole affair poisoned management-labor relations for years to come.

REFERENCES

1. B. Aplet, *The Corporation: A Centennial Biography of United States Steel Corporation, 1901–2001*. Cathedral Publishing, University of Pittsburgh, 2000
2. K. Warren, *Bethlehem Steel*, University of Pittsburgh Press, 2008

Selected References

J.F. Hogan, *The 1937 Chicago Steel Strike: Blood on the Prairie*, The History Press, 2014

T.J. Misa, *A Nation of Steel: The Making of Modern America 1865–1925*, The Johns Hopkins University Press, 1995

D. Nasaw, *Andrew Carnegie*, Penguin Group, 2006

R. Raymond, *Out of the Fiery Furnace: The Impact of Metals on the History of Mankind*. The Pennsylvania State University Press, 1986

M. Reutter, *Sparrow Point*, Summit Books, 1988

D. Rudacille, *Roots of Steel: Boom and Bust in an American Mill Town*, Pantheon Books, 2010

B.C. Stoddard, *STEEL: From Mine to Mill, The Metal that Made America*, Zenith Press, 2015

K. Warren, *Big Steel: The First Century of the United States Steel Corporation, 1901–2001*, University of Pittsburgh Press, 2008

K. Warren, *Bethlehem Steel*, University of Pittsburgh Press, 2008

Chapter Twelve

The Decline of the Integrated Steel Industry

Strike of 1959

The steel strike of 1959 was a watershed moment in the history of the U.S. integrated steel industry. The strike itself was not the main event, but it exposed industrywide problems and failures that had existed for many years. Among them were poor labor relations, delayed modernization, an unbalanced product mix, and inconvenient mill locations in relation to the market. Other problems included steady price increases to cover union contracts rather than productivity improvements, pollution that had to be controlled by the Environmental Protection Agency (EPA) rulings, and management staff who were paid the highest incomes of any industry.

In the early 1950s, David J. McDonald replaced Philip Murray, longtime president of the steelworkers union. McDonald was expected to be a moderate negotiator due to his limited experience in the steel union (**FIG. 12.1**). However, he had a bitter rivalry with Walter Reuther of the auto union, who had been elected head of the Congress of Industrial Organizations (CIO) after Murray. In subsequent negotiations, McDonald tried to keep the steelworkers ahead of the autoworkers. By 1959, steelworkers were earning roughly 38 cents per hour more than autoworkers, approximately 15% more.

Even so, McDonald never had the same support from union members that Murray had, so he always overreached in his attempt to keep them loyal. A serious strike took place in 1956, and workers were off the job for 35 days. The final settlement included Clause 2b, for Bethlehem only, which gave the union

FIG. 12.1 David J. McDonald, president of the steelworkers union during the strike of 1959. Source: www.usw.org.

a guarantee that new technology would not reduce the number of employees. This would be the source of labor problems at Bethlehem for years to come.

The 1959 strike was the longest in steel history at 116 days. Grievances included wages, pensions, and paid holidays. Bethlehem fought unsuccessfully to remove Clause 2b, and the strike became a pivotal point in the history of the U.S. steel industry. As the strike continued, customers became seriously concerned about supply sources, and their only recourse was imported steel. The United States had been a net exporter of steel during the 20th century, but in 1959, imports exceeded exports for the first time, with total imports reaching 4.7 million tons.

Technology and Imports

Due to the boom within the integrated steel industry during the 1950s and the forecast for increasing consumption in the future, the industry increased spending to upgrade existing plants and build new facilities. However, much of the new investment went into equipment that quickly became obsolete. As reported by Mark Reutter in his book *Making Steel: Sparrows Point and the Rise and Ruin of American Industrial Might*, "They relied on inflating the price of steel to keep up quarterly dividends and they let the situation deteriorate to the point where their mills were out of date" (Ref 1).

From 1957 to 1962, steel use declined 8% per capita, although the total market was still growing slowly. At the same time, the steel industry was trying to upgrade and expand capacity. But the adoption of modern technology was slow, as was evident when Bethlehem built a new open hearth shop in 1958 at Sparrows Point for $200 million. Four years earlier, a small steel company in Michigan had installed the first basic oxygen furnace (BOF), and the Jones and Laughlin Steel Company (J&L) installed these furnaces in 1959. The BOF process was invented in Austria after World War II. Basic oxygen furnace vessels were similar to the earlier Bessemer converters, except they used oxygen instead of air to remove the carbon from cast iron. A heat of steel could be processed in less than an hour, compared to six or eight hours for the open hearth furnaces, required fewer workers, and produced a superior product. New plants being built in Europe and Japan were using the new BOFs.

There were 21 BOFs in operation before the United States Steel Corporation (USS) built its first two at the Duquesne plant in Pittsburgh in 1963; there were 29 units in operation when Bethlehem Steel installed its first two at Sparrows Point in 1964, replacing open hearths that were 40 years old. That same year, the company built two BOFs at Lackawanna, which produced 2.5 million tons per year. The United States produced only 17% of its steel using BOFs in 1965, while Japan produced 55% of theirs. In 1970, this ratio was 48% for the United States and 79% for Japan.

Reluctance to accept new technology extended to the continuous casting process, another European development after World War II. Steel from the BOF was teemed into a machine that channeled the metal through a container, where it solidified and was prepared for the rolling mills. This process replaced the casting of metal into large molds and reducing them by forging or rolling in preparation for the finishing mills.

In 1961, Bethlehem Steel built a central research laboratory to compete with USS, which had built its own lab in 1957 (**FIG. 12.2**). Bethlehem built a huge campus on 1000 acres. The $35 million project—eight buildings and a reflecting pool—was not complete until 1969. It accommodated over 1000 research engineers and support personnel, and it was named Homer Research Laboratory after Bethlehem's chairman, Arthur B. Homer.

From 1960 to 1970, imports increased from 8% to 17% of total U.S. steel consumption. Most of the imported steel came from Japan, which had increased production from 25 to 100 million tons annually. The increase came from nine new plants located at tidewater ports for ease of importing raw materials and exporting finished product.

In 1964, Bethlehem decided to build a new mill at Burns Harbor, Indiana, featuring a new production layout and state-of-the-art equipment. It became the most efficient in the country. The first phase of construction was limited to

FIG. 12.2 Bethlehem Steel's Homer Research Laboratory. Source: Lehigh University.

finishing operations, and ingots were shipped in from Johnstown and Lackawanna. Blast furnaces and BOFs were added later to make it a fully integrated plant.

The first integrated steel company to be affected by the new market with increasing imports was J&L. They sold an interest in the company in 1968 to a Texas conglomerate called Ling-Temco-Vought (LTV). It had no experience in owning steel mills. A few years later, LTV took over total ownership of J&L and called it the LTV Steel Division. Thus, J&L became the first casualty in the decline of the integrated steel industry. The next company to disappear from the industry was YS&T. It closed the Campbell plant in Youngstown in the late 1970s and terminated 5000 employees. In 1979, it sold all remaining plants to LTV and joined J&L in the LTV Steel Division.

Though the winds of change had turned, both USS and Bethlehem built new corporate headquarters in the early 1970s. USS was first, constructing a $50 million, 64-story office in downtown Pittsburgh (**FIG. 12.3**).

FIG. 12.3 U.S. Steel corporate headquarters in downtown Pittsburgh.

In 1972, Bethlehem built Martin Tower in Bethlehem, Pennsylvania, named for its chairman, Edmund Martin. It was 21 stories high and cost $35 million (**FIG. 12.4**). The same year, Bethlehem had peak production of 16 million tons of finished product and earnings of $200 million. Employment reached 120,000 in the mid 1970s, but 20 years later only 20,000 workers remained. As written by Thomas Misa in *A Nation of Steel,* "The success of the industry through the 1960s led to the complacence at all levels of the largest companies. One tragic mistake was the flagrant disregard of users' needs and the stagnation in technical capabilities that led directly to industrial decline" (Ref 2).

In 1979, USS closed its Youngstown plant, a plant with open hearth furnaces, finishing mills dating back to 1900, and a sheet mill installed in 1935. It had become obvious that the mills were too outdated to compete in the new market conditions. In his book *Big Steel*, Kenneth Warren wrote, "Though still a major factor in world steel and unrivaled in size in the domestic U.S. industry,

FIG. 12.4 Martin Tower, headquarters of Bethlehem Steel in Bethlehem, Pennsylvania.

now more clearly than ever US Steel was an ailing giant leading a sickening industry. The 1970s were to bring conditions of unprecedented severity" (Ref 3).

In the late 1960s, USS bought a large tract of land near the city of Baytown, Texas, to build a new mill to make plate for the Southern market. This was a mill without blast furnaces and BOFs; it instead used electric furnaces to melt scrap and continuous casters to make slabs for the rolling mills. This concept was later called the minimill. This minimill was the third such mill in the United States; it was preceded by two small mills, North Star Steel in Minneapolis and a Nucor-built mill in Darling, South Carolina, that produced only small bars and shapes. It was another 16 years before a company attempted to build a plant to produce slabs for rolling into light plate and sheet. By that time, USS had already closed the Baytown works due to market conditions. It was later sold to Jindal Steel & Power Ltd., an Indian company.

Bethlehem finally built a continuous caster at Burns Harbor in 1975, after abandoning its first attempt to continuous cast slabs in 1969 at the Johnstown plant. At this time, the Japanese were continuous casting 46% of their steel capacity, compared with 11% for the U.S. integrated steel industry. Foreign imports held steady at 15% during the 1970s, but increased to 22% throughout the next decade.

In 1977, Bethlehem reported a loss of nearly $500 million and began closing parts of Johnstown and Lackawanna. The company terminated 2500 white-collar workers that year, 800 of them from the Homer Research Laboratory and Martin Tower facilities built just seven years earlier. In *Bethlehem Steel,* Kenneth Warren wrote, “Bethlehem Steel reached a turning point in 1977, a year when the momentum for expansion ran out and after which retrenchment and closure became dominant,” (Ref 4).

During the recession of 1982, total steel production fell from 120 million tons to just 75 million in the span of one year, and imports accounted for 25% of the market for the first time. USS operated at 36% of capacity, and Bethlehem lost $1.5 billion, with $900,000 of that from the cost of closing most of Lackawanna. By the end of 1983, employment in the Pittsburgh plant decreased from 27,000 to 9000. Bethlehem finally closed the open hearth shop at Sparrows Point, Maryland, which they had built as their hope for the future in 1958, when this type of shop had already been obsolete for 20 years.

Between 1982 and 1987, Bethlehem lost another $1 billion and cut its workforce in half, yet it invested $2 billion in capital expenditures. In 1986, the company reduced staff at the research laboratory and sold five of its eight buildings to Lehigh University. A few years later, Bethlehem sold the remaining buildings to Lehigh, ending the grand attempt to restore the company through basic research results. Over time, the company also reduced staff at Martin Tower; the building was vacant by 2007. It stands empty today and is contaminated with asbestos. By the end of the 1980s, Bethlehem found that it was hard to break the old culture with a board comprised of inside executives who continued to authorize projects that did not help restore profitability. The practice of spreading new investments over many plants instead of fully modernizing the Burns Harbor and Sparrows Point facilities wasted the investment funds.

Devastation in Pittsburgh

Closings during the 1980s hit the Monongahela River Valley near Pittsburgh especially hard. USS closed steelmaking at its Clairton Works in 1984, leaving only the coke-making operations functional. That same year, USS also closed Duquesne Works. In 1986, it closed Homestead Works and sold Christy Park Works. In 1983, the company’s Monroe Research Center had 3500 workers, but by 1991 the head count was reduced to 500, and the facility later closed. The only remaining USS plants in Pittsburgh were the Edgar Thompson Works, the finishing plant at Irvin, and the coke-making works at Clairton. Additionally, plants owned by J&L were closed in 1985. Thus, Pittsburgh—the heart of the integrated steel industry for a hundred years—was now a minor player in domestic steelmaking.

To offset the decline in steelmaking, USS bought the Marathon Oil Company in 1982. The company added to its oil holdings a few years later, purchasing Texas Oil and Gas Corporation. In 1986, it renamed the parent company USX, with USS as a subsidiary. These moves garnered a lot of criticism within the union and broader business world for using funds that could have gone into steel investments. However, Marathon Oil and Texas Oil and Gas earned profits over the next 20 years that supported the steel operations and provided funds to upgrade and purchase more efficient plants from bankrupt companies. USX divested its oil holdings in 2001 and changed its name back to USS.

Other Widespread Closures

Republic Steel, the next casualty of the steel decline, was sold to LTV in 1984. LTV now owned three of the top Little Steel companies in the country, all with histories going back to 1900. With its heavy investment in companies in distress and no knowledge of the steel industry, LTV went bankrupt in 1986. After recovering from this bankruptcy, it went bankrupt several more times, finally closing all facilities in early 2000. Remaining assets were bought by Wilbur Ross, a fund manager who specialized in buying companies at bargain prices and attempting to revive them for resale.

USS closed its Fairless Works in 1991. That same year it also closed the Baytown plant, which was only 20 years old. From 1980 to 1998, production decreased from 17 million tons to 11 million and the number of employees went from 150,000 to 20,000. By 1995, remaining major operations of USS included the Edgar Thompson works, the Clairton coke plant, the Irvin finishing mill in Pittsburgh, the Gary works, and its operation in Fairfield, Alabama.

A study conducted by an outside firm of consultants hired by Bethlehem concluded, "It could be argued that the company is in a state of slow liquidation, consuming its short and long term assets to fund operating losses." Production had declined from 19 million tons per year to 10 million from 1980 to 1992. By 1982, Bethlehem had closed four of its plants—Bethlehem, Steelton, Johnstown, and Lackawanna—and lost over $1 billion. Bethlehem had just 15,000 employees by 1999. However, imported steel was not the only factor in the decline of the integrated steel industry in the 1990s. Another was the inroads being made by the new minimill industry.

Minimills

The minimill industry, which had begun by making lower grade bars, angles, and reinforcing rods for concrete in 1970, had advanced to producing quality sheet and structural beams by the 1990s. This new industry was a major factor in Bethlehem's decline. The company's production decreased to nine million tons while Nucor, the minimill company, produced 10 million tons. After nearly 80 years, Bethlehem was no longer the second-largest steel producer in the country.

The year 1998 was a peak year in steel consumption in the United States at 140 million tons. However, imports accounted for 40 million tons, and the 30 million tons produced by the minimills represented half of all domestic steel used that year. Bethlehem built a new cold rolling mill at Sparrows Point at a cost of $600,000. When it went onstream, the market was falling and there was a worldwide excess of cold rolled capacity. This mill was never fully successful, and it was the last major investment the company would make.

National Steel Corporation, one of the small steel companies that had been successful from its formation in 1929 until the 1980s, began to lose momentum as a result of poor management. In 1994, the board fired the CEO and all the vice presidents and replaced them by recruiting a new team from the Gary works of USS. With the deteriorating markets, increasing imports, and competition from minimills, the new group of executives soon left the company. National Steel filed for bankruptcy in 2002 and was bought by USS for $1 billion.

In 2001, Bethlehem suddenly demoted its chairman after only one and a half years in office and hired a turnaround specialist to run the company. He placed Bethlehem into bankruptcy, and the company lost $1.9 billion that year. Out of the destruction, Wilbur Ross purchased the remaining assets in bankruptcy court and added them to his holdings in International Steel Group (ISG) (**FIG. 12.5**). Bethlehem Steel, an icon of American industry that had made everything from wire to armor plate to large forgings, built battle ships, aircraft carriers, cruisers, destroyers, landing ships, and freighters, as well as constructed the Empire State Building and the Golden Gate Bridge, ceased to exist on May 2, 2002.

Ross's holdings grew to include the LTV plants (the former J&L, YS&T, and Republic Steel) and the Burns Harbor and Sparrows Point works of Bethlehem Steel. He improved operations by canceling the union contracts in bankruptcy and terminating thousands of employees. By reducing the labor force and revoking the benefits the union had won during the previous 50 years, these plants became profitable once again. After several years, Ross sold ISG to Mittal Steel, a company whose largest shareholder was Lakshmi Mittal (**FIG. 12.6**).

Mittal

Mittal is an Indian company with worldwide holdings in steel production, headquartered in Luxembourg (**FIG. 12.7**). It would later acquire Arcelor Steel, the largest steel company in Europe, and change the name to ArcelorMittal. With ownership of the integrated mills from ISG, ArcelorMittal became the largest

FIG. 12.5 Wilbur Ross, a turnaround specialist for bankrupt companies.

FIG. 12.6 Lakshmi Mittal, chairman and CEO of ArcelorMittal. Source: Wikimedia Common.

FIG. 12.7 ArcelorMittal headquarters in Luxembourg.

steel producer in the United States at 22 million tons capacity. It is also the largest steel producer in the world, with 100 million tons global capacity. After 100 years, the biggest integrated steel producer in the United States now has overseas ownership.

The fate of Bethlehem, along with Youngstown Sheet and Tube, J&L, Republic Steel, and many other, smaller integrated steel companies, destroyed most of the remaining integrated steel companies in the United States. These companies left tens of thousands of employees with no jobs, no health insurance, pension funds that had to be taken over by the federal government, abandoned plants, and decimated cities; mill cities such as Sparrows Point, Bethlehem, Johnstown, Pittsburgh, Lackawanna, Youngstown, Gary, and South Chicago lost up to half of their populations. The devastation was epic, and the effects continue to ripple far into the future.

REFERENCES

1. M. Reutter, *Sparrows Point*, Summit Books, 1988
2. T.J. Misa, *A Nation of Steel: The Making of Modern America 1865–1925*, The Johns Hopkins University Press, 1995

3. K. Warren, *Big Steel: The First Century of the United States Steel Corporation, 1901–2001*, University of Pittsburgh Press, 2001
4. K. Warren, *Bethlehem Steel*, University of Pittsburgh Press, 2008

Selected References

B. Aplet, *The Corporation: A Centennial Biography of United States Steel Corporation, 1901–2001*, Cathedral Publishing, University of Pittsburgh, 2000

J.P. Hoerr, *And the Wolf Finally Came: The Decline of the American Steel Industry*, University of Pittsburgh Press, 1988

J.F. Hogan, *The 1937 Chicago Steel Strike: Blood on the Prairie*, The History Press, 2014

D. Rudacille, *Roots of Steel: Boom and Bust in an American Mill Town*, Pantheon Books, 2010

J. Strohmeyer, *Crisis in Bethlehem: Big Steel's Struggle to Survive*, Adler & Adler, 1986

P.A. Tiffany, *The Decline of American Steel: How Management, Labor, and Government Went Wrong*, Oxford University Press, 1988

Chapter Thirteen

The Steel Minimills

There was little change in the steelmaking industry from 1900 to 1960. The business was dominated by a few very large integrated companies led by U.S. Steel Corporation.

Steel minimills came on the scene in the late 1960s and in the 1970s. Besides being much smaller than the big mills, they featured three areas of technology that set them apart: Only scrap was used in the charge for the melting furnaces; electric arc furnaces (EAF) were used for melting; and steel was solidified into billets using continuous casting, which fed the billets directly into the rolling mills. Scrap had been melted in electric arc furnaces since the early part of the 20th century, but this practice was used only for making tool steels, high alloy steels, and stainless steels where scrap had to be high quality to meet specifications. Scrap was also used in large quantities in the open hearth process to make ordinary carbon steels and engineering alloy steels.

During the 1960s, big steel companies were changing production methods, moving to the new basic oxygen converter that removed carbon from the molten metal delivered from blast furnaces. This reduced the scrap market to very low levels with low prices, giving minimills further advantage against Big Steel, with its capital requirements for blast furnaces and basic oxygen converters. At the same time, Big Steel was adjusting to conversion to continuous casting. All of these factors opened the door for minimills to make small bars and shapes that did not need to meet the specifications of higher quality steel.

The First Minimill

The first continuous caster was installed in 1959 at the Premium Steel Company in Alberta, Canada. The work was done under Gerald Heffernan, a metallurgical

FIG. 13.1 Gerald Heffernan built the first minimill in North America, in Ontario, Canada, and the second minimill, in St. Paul, Minnesota. Source: www.amm.com.

engineer from the University of Toronto. Only a portion of production was put through the caster; the rest was teamed into ingots, which had to be rolled or forged into billets. Stelco of Ontario, Canada, purchased Premium Steel in 1963 to gain access to the growing market in the West.

With his share of the sale and other financial backing, Heffernan built the first minimill in North America in 1964, called Lake Ontario Steel Company (Lasco). Lasco was the first steel plant to put all hot metal through a continuous caster. Heffernan adopted a new management philosophy, developing training for many different job requirements with a high level of attention to safety. This management style became standard for the future minimill industry (**FIG. 13.1**).

The second plant built in North America was North Star Steel Company, in St. Paul, Minnesota, also engineered by Heffernan. North Star was financed by Cargill with minority stockholders (including Heffernan). It went onstream in 1968, producing 200,000 tons per year in bars and shapes. One potential customer was Vulcraft in Norfolk, Nebraska, a company that assembled small bars and angle shapes into roofing beams for shopping malls and big box buildings. Kenneth Iverson, Vulcraft's president, visited North Star to discuss buying steel, but came away determined to build his own minimill to supply five assembly plants (**FIG. 13.2**).

The parent company, Nuclear Corp (Nucor), selected a site in Darlington, South Carolina, near one of the fabrication plants. This minimill became the

FIG. 13.2 F. Kenneth Iverson, CEO of Nucor, led the minimill industry for the first 25 years. Source: www.aimehq.org.

third in North America and the second in the United States, with a capacity of 200,000 tons per year. Startup difficulties with local workers, combined with Heffernan's lack of involvement, nearly bankrupted the company. Even so, it went into full production in 1971 (**FIG. 13.3**). This plant became so successful that

FIG. 13.3 Nucor's minimill in Darlington, South Carolina, had initial capacity of 200,000 tons/year and originally produced long products. Circa 1969. Source: www.nucor.com

Nucor built two more minimills near its assembly plants in Norfolk and Jewett, Texas. Most of the minimills were located in small towns in rural areas.

Heffernan's Third Minimill

When Cargill bought out the minority shareholders of North Star Steel Company in 1972, Heffernan sold his shares and founded Co-Steel. He partnered with Texas Enterprise to build his third minimill, Chaparral Steel Company in Midlothian, Texas. Chaparral was equipped with the latest technology in EAFs and continuous casters and had a capacity of 200,000 tons per year, common with these first minimills. With constant upgrading of equipment and process improvements, Chaparral produced one million tons per year in 10 years and two million tons per year after another 20 years.

Through the leadership of Iverson and Heffernan, the minimill industry developed into 30 plants in the United States and eight in Canada during the 1970s. Typical of these additional minimills was Auburn Steel Company in Auburn, New York, built in 1974 with capacity of 400,000 tons per year. As with most of the new minimills, Auburn did not have an in-house market, as did Nucor with its Vulcraft Division and Chaparral with the oil industry. Instead, the mills were subject to the business cycles of the steel industry, and most of them were eventually acquired by bigger and more financially secure corporations. Nucor purchased Auburn Steel and at the same time built a Vulcraft plant nearby to use the output. In 1975, minimills had a combined output of six million tons—a small percentage of total steel production, which was more than 100 million tons that year.

The First Minimill for Flat Products

A huge breakthrough in the minimill industry was Iverson's new plant dedicated to producing flat products for Nucor. This product line required the highest quality steel, which was still the mainstay of the integrated steel industry. Big Steel thought this area could not be invaded by companies that relied on scrap for feedstock. The plant was built in 1988 at Crawfordsville, Indiana, and the heart of the process was a new continuous casting machine built by Siemens in Germany.

The continuous casting machine enabled Nucor to cast slabs 51 mm (2 in.) thick by 1320 mm (52 in.) wide, ready to enter the rolling mill to produce hot coil bands for cold rolling to the sheet product used by the automotive, appliance, and other industries. The Crawfordsville plant produced two million tons per year and cost $1 billion to build. This was the beginning of multimillion-ton plants, although the term minimills remained in use.

Steel Beams and Railroad Rails

At about the same time, Nucor teamed with a Japanese company, Yamato Steel Co., to produce structural products, wide flanged beams, and railroad rails in a new mill at Blytheville, Arkansas, with capacity of 2.8 million tons per year. Here again was a breakthrough that had a great impact on the integrated steel industry.

The sheet mill at Crawfordsville was immediately so successful that Nucor built a second sheet mill at Hickman, Iowa, with capacity of two million tons per year. This was the start of minimills' production of flat products and structural products that would make significant inroads into the primary business of integrated steel. For 25 years, Nucor led the industry in technology, production, management philosophy, and market penetration. Nucor's CEO, F. Kenneth Iverson, was the most important individual in the development of the minimill industry in the United States.

Iverson Leads Industry

F. Kenneth Iverson was born and raised in Donners Grove, Illinois. He attended Northwestern University in 1943 and served in the Navy from 1944 until 1946. He returned to college at Cornell University, where he received a bachelor's degree in aeronautical engineering. Iverson continued his engineering education with a master's in mechanical engineering at Purdue University. His first position was with International Harvester Company, and he later worked at Illium Corp., Indiana Steel Products, Cannon Muskegon, and Coast Metals. Iverson was a member of ASM International and AIME. He received honorary doctorates from the University of Nebraska and Purdue University. Nucor established the Iverson Chair at the University of Missouri in 2006 in his honor.

Iverson was inducted into the National Academy of Science, Engineering Division, in 1994 and was awarded the National Medal of Technology, presented by President George H.W. Bush. Iverson retired in 1996 for health reasons and died in 2002 at age 76.

Growth of the Minimill Industry

In the 20 years between 1988 and 2008, the minimill industry added 25 million tons annual capacity. This was equal to 25% of the total U.S. steel market. Competition for raw materials and markets led to consolidation of the industry, resulting in just three major minimill producers in the United States.

Nucor selected Dan DiMicco as its new CEO in 2000. He continued the momentum that had been established by Iverson. Nucor built a new continuous

FIG. 13.4 Nucor installed a Castrip continuous casting line at its plant in Hickman, Arizona. Source: www.castrip.com.

casting machine at its plant in Crawfordsville, Indiana, and a second one in Hickman, Arkansas, to produce what it calls "Castrip" (**FIG. 13.4**).

In the Castrip process, molten metal enters between two counter-rotating rolls, where it solidifies into a thin sheet (**FIG. 13.5**). This sheet is hot rolled with fewer passes than a long stand of rolls, producing a finished product with less energy, space, and time.

The Castrip concept was adopted at the Nucor plant that manufactured wide flanged structural beams and railroad rails. The continuous casting machines were rebuilt to cast a beam or rail near finished size (**FIG. 13.6**). This new shape, called "near-net shape," could then be rolled to finished size without

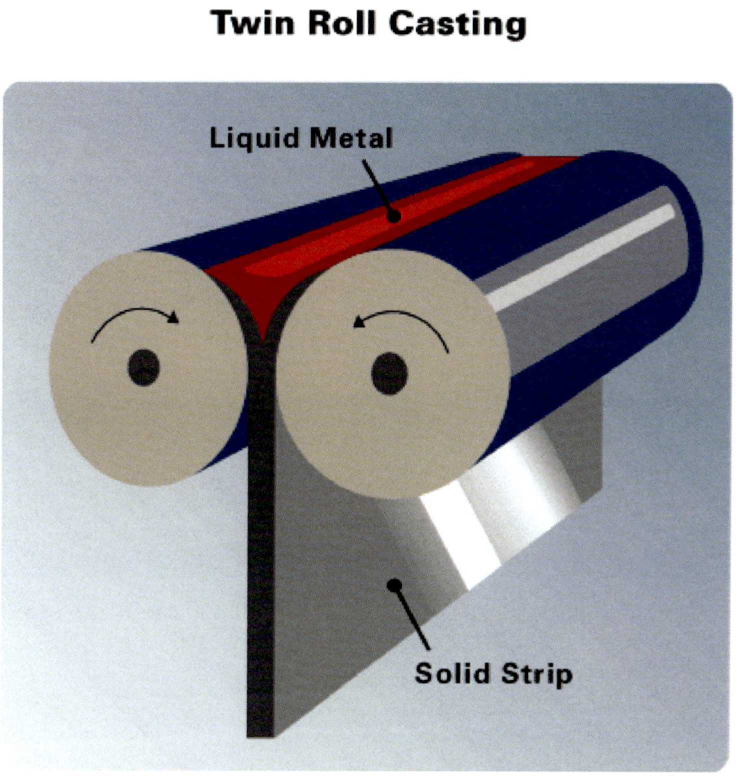

FIG. 13.5 Twin roll casting uses two copper water-cooled, counter-rotating rolls. The Castrip process produces a thin strip that requires minimal rolling to achieve finished size. Source: www.castrip.com.

the long train of rolls. The minimill industry had to follow Nucor's lead in this new technology to be competitive, and the large integrated steel mills were faced with another problem in maintaining sales in their best markets.

North Star Steel Company became the second largest minimill producer in the United States by 1997. The company moved beyond St. Paul, Minnesota, buying a mill in Wilton, Iowa, that went onstream in 1974 with 150,000 tons per year capacity. In 1977, it built a minimill at Duluth, Minnesota, to make parts for grinding mills at iron mines. In 1980, North Star built a minimill at Monroe, Michigan, to manufacture higher quality steel bars for the automotive industry. In 1983, the company purchased a mill at Beaumont, Texas, with a 500,000 ton capacity to make rod and wire products. They also bought a minimill from Hunt Steel Company in Youngstown, Ohio, in 1985. North Star increased capacity of the original St. Paul plant to 650,000 tons in 1993. In 1996, the company built a minimill in Kingman, Arizona, with 500,000 tons of capacity to make rod and

FIG. 13.6 The ability to make large flanged beams created a new market for the minimills. Source: www.steeldynamics.com.

wire for the West Coast market. North Star entered the flat products market in 1996 by partnering with BHP Steel, an Australian company. This plant, located in Northwest Ohio, had capacity of 1.5 million tons per year. North Star now had an annual steel capacity of 5 million tons per year.

In 1993, several executives at Nucor left the company to start their own minimill company, Steel Dynamics Inc. (SDI). Headquartered in Fort Wayne, Indiana, the company built its first minimill in Butler, Indiana, to produce flat products; the minimill came onstream in 1996. They entered the wide flange and railroad rail markets with a minimill in Columbia City, Indiana, at a cost of $315 million. Welding facilities were installed at this plant to join rail steel into lengths up to 1600 feet—a major advantage over the other rail mills.

Minimill Consolidation

After the year 2000, corporations grew through acquisitions, because many privately held mills found it difficult to compete for recycled steel and for markets where they were still producing only small bars and shapes. This trend was led by Nucor when they purchased Birmingham Steel Corporation in 2002, acquiring minimills in Alabama, Mississippi, Illinois, and Washington. Nucor then purchased the Kingman, Arizona, rod and wire mill facility from North Star in 2004. Later, it acquired Harris Steel in Canada for $1 billion and David J. Joseph, a major recycler of feedstock for minimills, for $1.4 billion. It became vital for the bigger minimill companies to enter the recycling industry to ensure scrap supply.

Nucor bought a cold rolling plant located at Sparrow's Point, Maryland, which had been part of the now-bankrupt Bethlehem Steel Company. Nucor then acquired the Gallatin Steel minimill located in Kentucky for $770 million. Nucor's production capacity reached 28 million tons per year, exceeding that of U.S. Steel. It was not only the biggest minimill, at 50% of the market, but the largest steel company, with 25% of all steelmaking in the United States.

A new company came on the scene in 1999 and began buying minimills—Gerdau, a name unfamiliar to the American public. Gerdau, with headquarters in Brazil, had holdings of both minimills and integrated mills in South America, Central America, and Mexico. Its first purchase was Florida Steel Corporation, a minimill making concrete reinforcing rods with four plants in Florida, Tennessee, and North Carolina. Gerdau bought four mills from North Star in 2004. Over the next decade, the company acquired 10 more minimills, for a total of 18 in 12 states and Canada. Most of these mills were the older ones making the long products of bars and shapes. Gerdau's most important acquisition was Chaparral Steel Company, the minimill built and improved over the years by Gerald Heffernan. Chaparral enabled Gerdau to enter the wide-flanged beam and rail markets. Without building a single minimill, Gerdau became the third largest minimill producer in the United States. In addition, with its holdings, Gerdau approached Nucor in production capacity and sales.

Steel Dynamics Inc. (SDI) bought Roanoke Electric Steel Corporation in 2006, its first acquisition, which included Steel of West Virginia Inc. The company also bought into the manufactured beam market, acquiring New Millennium Building Systems. SDI's strategy followed that of Nucor—having an in-house market for long products. SDI made its biggest acquisition in 2014 when it purchased all the assets of a large sheet product mill in Columbus, Mississippi, for $1.63 billion. Severstal, a Russian company, built this minimill in 2007. It featured the latest technology and a capacity of 3.3 tons per year. SDI's production capacity in sheet, structural beams, rail, and long products was now 11 million tons per year, making it the second largest minimill company in the United States This was accomplished in just 18 years since opening its first minimill.

North Star, the first minimill in the United States, was still second in production in 1997, with a capacity of 5 million tons per year. The company upgraded its mills, added new ones, and entered the flat steel product markets. In the early 2000s, Cargill (owner of North Star) decided to leave the steel industry, retaining only the flat product mill in Ohio they owned jointly with Bluestone (formerly BHP) of Australia. North Star sold its newest mill in Kingman, Arizona, to Nucor

FIG. 13.7 Nucor built a direct reduction plant to make iron units for its electric furnaces in St. James Parish, Louisiana. Source: Nucor presentation.

and four of its other mills to Gerdau, including the original one at St. Paul. North Star is now a minor player in the minimill industry.

New Processes for Making Iron

Availability of recycled steel for the minimills was a critical problem due to rapid industry growth after 1995. The price for high-grade scrap had increased from $30 per ton in 1970 to $125 in 1995, reaching $150 by 2005. The big producers, especially Nucor and Steel Dynamics, responded by buying national recycling companies to ensure supply, but they had no control over the price, which was set by the global market.

Next, iron units were sought for electric arc furnaces for use in new processes that converted iron ore by direct reduction—in other words, by removing oxygen without adding carbon. Other elements in the ore were then eliminated using slags in the electric furnace. Steel Dynamics installed a plant to make iron as a feedstock in their Butler, Indiana, plant. The latest development is a direct reduction plant built in Louisiana by Nucor (**FIG. 13.7**).

The introduction of flat products, flanged beams, and railroad rails, along with the consolidation of the industry into large corporations, completed a revolution in steelmaking in the United States, albeit a quiet one that occurred in small towns and rural areas out of sight of the general public. There is no Pittsburgh of minimills, and this revolution came at the expense of the large integrated steel mills—most large mills disappeared, with their best assets bought by U.S. Steel or ArcelorMittal, an India-based company with the world's largest capacity for steelmaking.

Selected References

F.K. Iveson, *Plain Talk*, John Wiley & Sons, Inc., 1998

M. Reutter, *Sparrows Point*, Summit Books, 1988

J. Stubbles, *The Minimill Story*, The Association of Iron & Steel Technology, 2006

K. Warren, *Big Steel: The First Century of the United States Steel Corporations, 1901–2001*, University of Pittsburgh Press, 2001

K. Warren, *Bethlehem Steel*, University of Pittsburgh Press, 2008

L. Chavez, *The Rise of Mini Steel Mills, The New York Times*, 23 Sept 1981, www.nytimes.com/1981/09/23/business/the-rise-of-mini-steel-mills.html

M. Metzger, F. Kenneth Iverson of Nucor: Man of Steel, *Inc.*, 1 April 1984, www.inc.com/magazine/19840401/6059.html

The North Star Steel Company History, *Funding Universe*, 1993, www.fundinguniverse.com/company-histories/north-star-steel-company-history/

The Evolution of America's Leading Steel Producer, Chapter 2: History, Nucor, 2015, http://www.nucor.com/story/chapter2/

Nucor Endows the F. Kenneth Iverson Steel Technology Chair at University of Missouri—Rolla 2006, Nucor Press Release, 25 May 2006, www.nucor.com/investor/news/print/?rid=860953

F. Kenneth Iverson, 76; Built Nucor Into No. 2 Steelmaker, *Los Angeles Times*, 20. April 2002, articles.latimes.com/2002/apr/20/local/me-ken20

Index

Kolibri 360 GmbH
Ferdinand-Jühlke-Straße 7
99095 Erfurt, Deutschland
produktsicherheit@kolibri360.de